T0188519

RECONFIGURABLE RADIO SYSTEMS

RECONFIGURABLE RADIO SYSTEMS

NETWORK ARCHITECTURES AND STANDARDS

Maria Stella Iacobucci
Telecom Italia HR Services, Italy

A John Wiley & Sons, Ltd., Publication

This edition first published 2013
© 2013 John Wiley & Sons, Ltd

Registered office
John Wiley & Sons Ltd, The Atrium, Southern Gate, Chichester, West Sussex, PO19 8SQ, United Kingdom

For details of our global editorial offices, for customer services and for information about how to apply for permission to reuse the copyright material in this book please see our website at www.wiley.com.

The right of the author to be identified as the author of this work has been asserted in accordance with the Copyright, Designs and Patents Act 1988.

All rights reserved. No part of this publication may be reproduced, stored in a retrieval system, or transmitted, in any form or by any means, electronic, mechanical, photocopying, recording or otherwise, except as permitted by the UK Copyright, Designs and Patents Act 1988, without the prior permission of the publisher.

Wiley also publishes its books in a variety of electronic formats. Some content that appears in print may not be available in electronic books.

Designations used by companies to distinguish their products are often claimed as trademarks. All brand names and product names used in this book are trade names, service marks, trademarks or registered trademarks of their respective owners. The publisher is not associated with any product or vendor mentioned in this book. This publication is designed to provide accurate and authoritative information in regard to the subject matter covered. It is sold on the understanding that the publisher is not engaged in rendering professional services. If professional advice or other expert assistance is required, the services of a competent professional should be sought.

MATLAB is a trademark of The MathWorks, Inc. and is used with permission. The MathWorks does not warrant the accuracy of the text or exercises in this book. This books use or discussion of MATLAB software or related products does not constitute endorsement or sponsorship by The MathWorks of a particular pedagogical approach or particular use of the MATLAB software.

Limit of Liability/Disclaimer of Warranty: While the publisher and author have used their best efforts in preparing this book, they make no representations or warranties with the respect to the accuracy or completeness of the contents of this book and specifically disclaim any implied warranties of merchantability or fitness for a particular purpose. It is sold on the understanding that the publisher is not engaged in rendering professional services and neither the publisher nor the author shall be liable for damages arising herefrom. If professional advice or other expert assistance is required, the services of a competent professional should be sought.

Library of Congress Cataloging-in-Publication Data applied for.

Hardback ISBN: 9781119969303

A catalogue record for this book is available from the British Library.

Typeset in 10/12pt Times by Aptara Inc., New Delhi, India
Printed and bound in Singapore by Markono Print Media Pte Ltd

Contents

Preface

Network evolution in the past decade involved the introduction of new access technologies, both fixed and wireless, using an IP backbone for all originating and terminating services. The evolution of the fixed access network mainly concerned the introduction of the optical fibre, with point-to-point or passive optical network (PON) architectures.

The next generation access network (NGAN) also includes radiomobile and wireless access technologies that, thanks to the adoption of advanced radio features, reach a maximum bit rate of hundreds of Mbps. The different access networks, which collect all originating and terminating services, are connected to an IP-based backbone, offering a transport service with quality of service (QoS).

The radio access technologies (RATs) are often managed by 'manually' configuring the radio parameters. Recently, some common radio resource management (CRRM) features have been introduced in the multi-RAT network in order to optimize the performances of the overall radio access network. The evolution of CRRM adds more intelligence to the access network, with self-organizing networks (SONs) and cognitive radio (CR). SONs are networks able to autoconfigure, self-manage and self-heal. This concept has been pushed by the industry forum of the next generation mobile network (NGMN) and SON features have been introduced in 3GPP standards starting from release 8.

All over the world a refarming process of the radiomobile spectrum is going on. In the near future, radiomobile frequencies will not be strictly associated to a technology, but will be used depending on user terminals, service profiles, traffic requests and network optimization.

With spectrum sensing cognitive radio terminals and networks, spectrum usage will be optimized among different radio access technologies. In this context, new scenarios are opening up different degrees of freedom: from the case of a licensed operator using cognitive radio inside its network to increasing the efficient use of radio resources, through the coordination of licensed operators for the spectrum usage, to a scenario where unlicensed cognitive radio terminals operate in times and zones where a licensed spectrum is underutilized. Cognitive radio also opens up a scenario where the spectrum resource can be managed in an hour-to-hour market for spectrum exchange.

In December 2010 the FCC gave a green light for the use of 'white spaces', that is the vacant spectrum in TV bands. White space transmissions are based on spectrum sensing and a geo-detecting database system to protect TV signals from interference. This is the first example of cognitive radio usage, which implies the deployment of the first wireless standard based on cognitive radio, IEEE 802.22. Full cognitive radio, in which every possible parameter observable by a wireless node or network is taken into account for adaptation and network optimization, evolves towards reconfigurable radio systems (RRSs), in both the network and terminal sides.

Starting from 2009, ETSI published deliverables on RRS and a series of recommendations for standardization. The ETSI RRS functional architecture proposes single operator and multioperator scenarios, with new entities for dynamic spectrum management, dynamic self-organizing network planning and management, joint radio resource management and configuration control. The new entities bidirectionally communicate through standard interfaces and with their corresponding parts in the radio terminal. On February 2009, the IEEE 1900.4 standard for reconfigurable radio systems was published. IEEE 1900.4 is a standard for architectural building blocks enabling network-device distributed decision making for optimized radio resource usage in heterogeneous wireless access networks.

From April 2009, the IEEE 1900.4 Working Group has been working on two projects: 1900.4a, an amendment to IEEE 1900.4 defining architecture and interfaces for dynamic spectrum access networks in white space frequency bands, and 1900.4.1, a standard for interfaces and protocols enabling distributed decision making for optimized radio resource usage in heterogeneous wireless networks.

The IEEE 1900.4 standard proposed architecture introduces new entities in both the terminal and network sides. In the network side, the new entities are the operator spectrum manager, the RAN measurement collector, the network reconfiguration manager and the RAN reconfiguration controller. All those blocks, except the operator spectrum manager, have pairs in the terminal side and standardized interfaces.

The book is structured as follows.

Chapter 1 presents an overview of the principal radiomobile and wireless systems, like GSM/GPRS/EDGE, UMTS/HSPA/HSPA+, LTE/LTE advanced, wireless LANs, wireless MANs and wireless PANs, and describes the performances, the network architectures and the radio access technologies.

In Chapter 2, the spectrum sensing cognitive radio and the related features at the transmitter and receiver sides are introduced. The techniques used at the physical layer to avoid interference to primary users, like spectrum sensing at the receiver side and adaptive modulation, coding and power control at the transmitter side, are described. At the end of the chapter full cognitive radio, more extensively covered in other chapters of the book, is introduced.

Chapter 3 introduces self-organizing networks and presents the self-organizing network features introduced in the 3GPP standard, like self-establishment, self-optimization and self-healing of the LTE access network.

Chapter 4 presents the concept of 'white space' and describes coexistence issues in TV bands. IEEE 802.22, the first standard based on cognitive radio, is described in terms of architecture and the principal features at the physical and MAC layers are shown. At the end of the chapter the IEEE 802.22.1 beaconing system for incumbent interference protection is introduced.

Chapter 5 presents the ETSI Functional Architecture (FA) for Reconfigurable Radio Systems (RRSs). Reconfigurable radio base stations and reconfigurable radio devices architectures are described, with related use cases. The concept of cognitive pilot channel, which conveys the necessary information from the network to the terminal in the cognition cycle, is introduced in both in-band and out-band possibilities, and related examples are given. The entities of the functional architecture in both the network and terminal sides together with the connecting interfaces are described.

Chapter 6 introduces the IEEE 1900.4 standard for reconfigurable radio systems. IEEE 1900.4 entities in both the network and terminal sides with the connecting interfaces are described and examples of procedures are given. The IEEE 1900.4a standard, which introduces new entities for dynamic spectrum access networks in white space frequency bands is presented, with its entities in both the network and terminal sides with connecting interfaces.

The last chapter of the book (Chapter 7) presents, without pretending to be exhaustive, the evolution from an exclusive usage of a spectrum regulated by individual licenses to the dynamic spectrum management and discusses how the systems described in the book enable dynamic spectrum management approaches.

Disclaimer

This text was prepared entirely on the author's personal time and with personal resources.

The author's affiliation with HRS Telecom Italia is provided for identification purposes only.

The opinions expressed in this text are those of the author and do not necessarily reflect the views of the author's affiliation company or of the standardization bodies.

Acknowledgements

My first acknowledgement is to my mentor, Prof. Maria Gabriella Di Benedetto. Some years ago, she taught me the method of scientific research and supported me during all of my PhD period. I would like to thank Domenico Di Giancristofaro and Giovanni Lofrumento for having read and commented on some parts of the book. I would also like to thank my colleagues from HRS for valuable discussions about some of the topics treated in the book: Giuliano Paris, Sandro Pileri, Italo Tobia, Giulio Di Vitantonio, Osvaldo Prosperi, Stefania Pace, Benvenuto Ioannucci and Santino Paciotti.

I appreciated the kind support provided by John Wiley & Sons, Ltd, in particular Tom Carter, Claire Foo, Sandra Grayson, Susan Barclay, Sophia Travis and Mark Hammond. I also appreciated the work of Patricia Bateson and Sharma Shalini during the editing process.

I thank my husband Silvio and my sons Raffaele and Maria because of the constant support they have given me during the writing of the book.

I will be grateful to all readers who open discussions about the topics of the book and will also appreciate comments and suggestions that could be considered in forthcoming editions of this book. The feedback is welcome to my email address mariastella.iacobucci@gmail.com. I will also be pleased to get in touch with readers through my LinkedIn account.

My last thought is not to a person, but to a beautiful historical city which on 6 April 2009 was destroyed by a tremendous earthquake: L'Aquila. This is the city where I was born, I grew up and where I actually live with my family. Its historical centre is now uninhabited, empty and lifeless. I hope to have the chance to see once more life inside the houses, the streets and the medieval churches.

With this hope, I wish you a pleasant reading.

List of Abbreviations

AASA	Aggregation aware spectrum assignment
ADSL	Asymmetric digital subscriber line
AGCH	Access grant channel
AICH	Acquisition indicator channel
AICPC	Acquisition indication CPC
AIP	Administrative incentive pricing
AMC	Adaptive modulation and coding
ANR	Automatic neighbour relation
AOA	Angle of arrival
AP	Access point
APAC	Asia Pacific
APT	Asia Pacific Telecommunity
ARQ	Automatic repeat request
ASA	Authorized shared access
ASN-GW	Access service network gateway
ATM	Asynchronous transfer mode
BCCH	Broadcast control channel
BCH	Broadcast channel
BER	Bit error rate
BLM_REQ	Bulk measurement request
BPSK	Binary phase shift keying
BS	Base station
BSC	Base station controller
BSIC	Base station identity code
BSS	Base station subsystem
BTS	Base transceiver station
BW	Bandwidth
CAC	Command and control
CAF	Cyclic autocorrelation function
CAP	Contention access period
CBP	Coexistence beacon protocol
CBS	Cognitive base station
CBSMC	CBS measurement collector
CBSRC	CBS reconfiguration controller

CBSRM	CBS reconfiguration manager
CC	Call control (Chapter 1)
CC	Channel contention (Chapter 4)
CC	Configuration control (Chapter 5)
CCB	Configuration control block
CCCH	Common control channel
CCH	Control channel
CCK	Complementary code keying
CCM	Configuration control module
CCN	Channel contention number
CCNCT	Channel contention number of credit token
CCS	Channellization code set
Cell-Id	Cell identifier
CEPT	European Conference of Postal and Telecommunications Administration
CF	Cognitive functionality
CFP	Contention free period
CGI	Cell global identity
CI	Cell identity
CIO	Cell individual offset
CITEL	Inter-American Telecommunication Commission
CM	Configuration manager
CMTR	Coordinated multipoint transmission and reception
CP	Control plane
CP	Cyclic prefix (Chapter 5)
CPC	Cognitive pilot channel
CPE	Customer premise equipment
CPS	Common part sublayer
CQI	Channel quality indicator
CR	Cognitive radio
CRC	Cyclic redundancy check
CRRM	Common radio resource management
CRS	Cognitive radio system
CS	Circuit switched (Chapter 1)
CS	Coding scheme (Chapter 1)
CSDB	Communications Society Standards Development Board
CSFB	Circuit switched fallback
CSG	Closed subscriber group
CSI	Channel state information
CSMA/CA	Carrier sense multiple access with collision avoidance
CT	Credit token
CTAP	Channel time allocation period
CTS	Clear to send
CUS	Collective use of spectrum
CW	Contention window
CWM	Composite wireless network
CWMP	CPE WAN management protocol

CWN	Composite wireless network
DBCPC	Downlink broadcast CPC channel
DCCH	Dedicated control channel
DCD	Downstream channel descriptor
DCF	Distributed coordination function
DCH	Dedicated channel
DFT	Discrete Fourier transform
DIFS	Distributed interframe space
DODCPC	Downlink on demand CPC
DPCCH	Dedicated physical control channel
DPCH	Dedicated physical channel
DPDCH	Dedicated physical data channel
DS	Downstream
DSA	Dynamic spectrum access (Chapter 7)
DSA	Dynamic spectrum allocation (Chapter 5)
D-SCH	Downlink shared channel
DSM	Dynamic spectrum management
DS-MAP	Downstream MAP
DSONPM	Dynamic self-organizing network planning management
DSSS	Direct sequence spread spectrum
DTCH	Dedicated traffic channel
DTV	Digital television
E-AGCH	Enhanced absolute grant channel
EBD	Eigenvalue-based detection
ECC	Electronic Communication Commission
ECGI	E-UTRAN cell global identity
E-DCH	Enhanced dedicated channel
EDGE	Enhanced data rates for GSM evolution
E-DPCCH	Enhanced dedicated physical control channel
E-DPDCH	Enhanced dedicated physical data channel
EGC	Equal gain combining
E-HICH	Enhanced HARQ indicator channel
EIRP	Effective isotropic radiated power
EM	Element manager
EMEA	Europe, Middle East and Africa
EMS	Element management system
eNB	Evolved Node B
EPC	Evolved packet core
EPS	Evolved packet system
E-RGCH	Enhanced relative grant channel
ES	Energy saving
ESS	Extended service set
ETSI	European Telecommunication Standard Institute
E-UTRAN	Evolved UTRAN
FA	Functional architecture
FACCH	Fast associated control channel

FACH	Forward access channel
FC	Flow controller
FC_ACK	Frame contention acknowledgement
FC_DST	Frame contention destination
FC_REL	Frame contention release
FC_REQ	Frame contention request
FC_RES	Frame contention response
FC_SRC	Frame contention source
FCC	Federal Communication Commission
FCCH	Frequency correction channel
FCH	Frame control header
FDD	Frequency division duplex
FDMA	Frequency division multiple access
FEC	Forward error correction
FEM	Front end module
FER	Frame error rate
FH	Frequency hopping
FIR	Finite impulse response
FR	Full rate
FS	Fixed station
FSK	Frequency shift keying
FSM	Frequency spectrum management
FT	Fourier transform
FTP	File transfer protocol
FTTB	Fibre to the building
FTTCab	Fibre to the cabinet
GERAN	GSM EDGE radio access network
GGSN	Gateway GPRS support node
GL	Geo-location
GMSK	Gaussian minimum shift keying
GPON	Gigabit passive optical network
GPRS	General packet radio service
GPS	Global positioning system
GSM	Global system for mobile communications
HARQ	Hybrid ARQ
HCS	Header check sequence
HeMS	Home eNB management system
HII	High interference indicator
HLR	Home location register
HO	Handover
HOM	Handover hysteresis margin
HPR	Hardware processing resource
HR	Half rate
HS-DPCCH	High speed dedicated physical control channel
HS-DSCH	High speed downlink shared channel
HSN	Hopping sequence number

HSPA	High speed packet access
HSS	Home subscriber server
HS-SCCH	High speed shared control channel
IBSS	Independent basic service set
ICH	Indicator channel
ICIC	Intercell interference coordination
Id	Identifier
IE	Information element
IEEE DySPAN-SC	IEEE Dynamic Spectrum Access Networks Standards Committee
IETF	Internet Engineering Task Force
IM	Identity management
IMS	IP multimedia subsystem
IMSI	International mobile subscriber identity
IMT	International Mobile Telecommunications
IP	Internet protocol
IR	Incremental redundancy
IRC	Interference rejection combining
ISM	Industrial, Scientific and Medical
ITU	International Telecommunication Union
JRRM	Joint radio resource management
KPI	Key performance indicator
LA	Location area
LAC	Location area code
LLC	Logical link control
LSA	Licensed shared access
LTE	Long term evolution
M2M	Machine to machine
MAC	Medium access control
MANET	Mobile ad hoc network
MB	Marginal benefit
MBFEM	Multiband front end module
MBMS	Multimedia Broadcast Multicast Services
MC	Management centre
MCCH	Multicast control channel
MCH	Multicast channel
MCS	Modulation and coding scheme
MGW	Media gateway
MIB	Management information base
MIB	Master information block
MIMO	Multiple input multiple output
ML	Maximum likelihood
MM	Mobility management
MME	Mobility management entity
MRC	Maximum ratio combining
MRC	Multiradio controller
MS	Mobile station

MSA	Maximum satisfaction algorithm
MSC	Mobile switching centre
MSDU	MAC service data unit
MSISDN	Mobile station ISDN number
MSK	Minimum shift keying
MTCH	Multicast traffic channel
NAS	Nonaccess stratum
NAV	Network allocation vector
NCI	Neighbour cell identity
NCMS	Network control and management system
NFC	Near field communication
NGAN	Next generation access network
NIST	National Institute of Standards and Technology
NM	Network manager
NMEA	National Marine Electronic Association
NMS	Network management system
NO	Network operator
NPD	Next-in-line protecting device
NR	Neighbour cell relation
NRM	Network reconfiguration manager
NRT	Neighbour relation table
O&M	Operation and maintenance
OAS	Organization of American States
ODFC	On demand frame contention
ODSC	On demand spectrum contention
OFDM	Orthogonal frequency division multiplexing
OFDMA	Orthogonal frequency division multiple access
OI	Overload indicator
OLT	Optical line termination
ONU	Optical network unit
OQPSK	Offset quadrature phase shift keying
OSA	Opportunistic spectrum access
OSM	Operator spectrum manager
OSS	Operational support system
OTA	Over the air
OVSF	Orthogonal variable spreading factor
PACCH	Packet associated control channel
PAD	Padding
PAGCH	Packet access grant channel
PAPR	Peak-to-average power ratio
PBCCH	Packet broadcast control channel
PBCH	Physical broadcast channel
PC	Point coordinator
PCCCH	Packet common control channel
PCCH	Paging control channel
PCCPCH	Primary common control physical channel

PCF	Point coordination function
PCF	Policy control function
PCFICH	Physical control format indicator channel
PCH	Paging channel
PCI	Physical cell identity
PCRF	Policy and charging resource function
PCS	Personal communications service
PCU	Packet control unit
PD	Protecting device
PDA	Personal digital assistant
PDCCH	Packet dedicated control channel (Chapter 1)
PDCCH	Physical downlink control channel (Chapter 1)
PDCP	Packet data convergence protocol
PDSCH	Physical downlink shared channel
PDTCH	Packet dedicated traffic channel
PDU	Packet data unit (Chapter 1)
PDU	Protocol data unit (Chapter 4)
PGW	Packet gateway
PHICH	Physical hybrid ARQ indicator channel
PHY	Physical layer
PICH	Paging indicator channel
PIFS	PCF interframe space
PLMN	Public land mobile network
PMCH	Physical multicast channel
PNC	Piconet controller
PNCH	Packet notification channel
PON	Passive optical network
PPCH	Packet paging channel
PPD	Primary protecting device
PRACH	Physical random access channel
PRM	Protocol reference model
PS	Packet switched (Chapter 1)
PS	Portable station (Chapter 4)
PSD	Power spectral density
PSK	Phase shift keying
PSS	Primary synchronization signal
PTCCH	Packet timing advance control channel
PTCH	Packet traffic channel
PTM-M	Point-to-multipoint multicast
PUCCH	Physical uplink control channel
PUSCH	Physical uplink shared channel
QAM	Quadrature amplitude modulation
QoS	Quality of service
QP	Quiet periods
QPSK	Quadrature phase shift keying
RA	Routing area

RACH	Random access channel
RACPC	Random access CPC
RAN	Radio access network
RAT	Radio access technology
RBS	Radio base station
RCM	Radio connection manager
RE	Reconfiguration entity
REG-REQ	Registration request
REG-RES	Registration response
RF	Radio frequency
RFB	Radio frequency block
RLC	Radio link control
RLC-PDU	RLC-packet data unit
RLF	Radio link failure
RMC	RAN measurement collector
RNC	Radio network controller
RNC-Id	Radio network control identifier
RNS	Radio network system
RNTPI	Relative narrowband transmission power indicator
RR	Radio regulations
RRC	Radio resource control
RRC	RAN reconfiguration controller
RRD	Reconfigurable radio device
RRM	Radio resource management
RS	Reference signal (Chapter 1)
RS	Resource sharing (Chapter 4)
RSRP	Reference signal received power
RSS	Received signal strength
RTS	Request to send
RTT	Round trip time
RVCC	Radio voice call continuity
SACCH	Standalone associated control channel
SAE GW	System architecture evolution gateway
SAE	System architecture evolution
SAP	Service access points
SC_MMF_EM	Self-configuration monitoring and management function element manager
SC_MMF_NM	Self-configuration monitoring and management function network manager
SCCPCH	Secondary common control physical channel
SCF	Spectral correlation function
SC-FDMA	Single carrier – frequency division multiple access
SCH	Superframe control header
SCH	Synchronization channel
SCW	Self-coexistence window
SDCCH	Standalone dedicated control channel

SDR	Software defined radio
SF	Spreading factor
SGSN	Serving GPRS support node
SGW	Serving gateway
SH_CO_F	Self-healing cell outage function
SH_MMF	Self-healing monitoring and management function
SH_MMF_EM	Self-healing monitoring and management function element manager
SH_MMF_NM	Self-healing monitoring and management function network manager
SIC	Successive interference cancellation
SIFS	Short interframe space
SIG	Special interest group
SIM	Subscriber identity module
SM	Session management (Chapter 1)
SM	Spectrum manager (Chapter 7)
SNMP	Simple network management protocol
SNR	Signal to noise ratio
SO_MMF_EM	Self-configuration monitoring and management function element manager
SO_MMF_NM	Self-configuration monitoring and management function network manager
SO-MMF	Self-optimization monitoring and management function
SON	Self-organizing network
SPD	Secondary protecting device
SS	Spectrum sensing
SSA	Spectrum sensing automation
SSS	Secondary synchronization signal
STM	Synchronous transfer mode
SW	Software
TACS	Total access communication system
TB	Transport block
TBS	Transport block size
TBTT	Target beacon transmission time
TC	Technical Committee
TCH	Traffic channel
TCI	Target cell identity
TDD	Time division duplex
TDMA	Time division multiple access
TE	Terminal equipment
TF	Transmission format
TMSI	Temporary Mobile Subscriber Identity
TMC	Terminal measurement collector
TOA	Time of arrival
TRC	Terminal reconfiguration controller
TRM	Terminal reconfiguration manager
TTG	Transmit/receive transition gap
TTI	Transmission time interval

TTT	Time to trigger
TVBD	Television band device
UCD	Upstream channel descriptor
UCS	Urgent coexistence situation
UE	User equipment
UMTS	Universal mobile telecommunication system
UP	User plane
US	Upstream
US-MAP	Upstream MAP
UTRAN	UMTS terrestrial radio access network
VLR	Visitor location register
VoIP	Voice over IP
WAVE	Wireless access in vehicular environments
WG	Working group
WLAN	Wireless local area network
WMAN	Wireless metropolitan area network
WPAN	Wireless personal area network
WRAN	Wireless regional area networks
WS	White space (Chapter 6)
WS	Wireless station (Chapter 1)
WSD	White space device
WSM	White space manager

1

The Multiradio Access Network

1.1 Introduction

Network evolution in the past decade regarded the introduction of new access technologies, both fixed and wireless, using an Internet protocol (IP) backbone for all originating and terminating services. The evolution of the fixed access network mainly concerns the introduction of the optical fibre, with point-to-point or passive optical network (PON) architectures.

Gigabit passive optical network (GPON) architectures deal with fibre optic deployment up to different points in the access network:

- Fibre to the cabinet (FTTCab), if the fibre stops at the cabinet,
- Fibre to the building (FTTB), if the fibre stops at the building, and
- Fibre to the home (FTTH), if the fibre stops at the customer's home.

Figure 1.1 shows FTTCab, FTTB and FTTH network architectures. Such architectures reach a downstream bit rate per user in the order of magnitude respectively up to 50 Mbps, up to 100 Mbps and up to 1 Gbps. The optical network is called passive because of the splitters, which repeat the input signal. The outgoing bandwidth of an optical line termination (OLT) is shared among many optical network units (ONUs), and in FTTCab and FTTB architectures the existing copper cable pair is used in the connection from the ONU up to the end users, with very high digital subscriber line (VDSL) transmissions. If the optical fibre reaches the home, the architecture is FTTH and the user will be provided with an optical modem called network termination (NT).

In the point-to-point architecture, there is one optical fibre connecting the end user to the central office, completely replacing the copper cable pair. In this case one fibre is dedicated to one user and therefore the provided bandwidth can be very high, even tens of Gbps. Figure 1.2 shows an example of the point-to-point fibre architecture in the access network.

The point-to-point architecture, handling much more fibre optics than GPON, requires more spaces in the central office and absorbs much more power. Because of that, most operators have chosen the GPON architecture for fixed access network evolution.

The next generation access network (NGAN) also includes radiomobile and wireless access technologies that, thanks to the adoption of advanced radio features, reach a maximum bit

Reconfigurable Radio Systems: Network Architectures and Standards, First Edition. Maria Stella Iacobucci.
© 2013 John Wiley & Sons, Ltd. Published 2013 by John Wiley & Sons, Ltd.

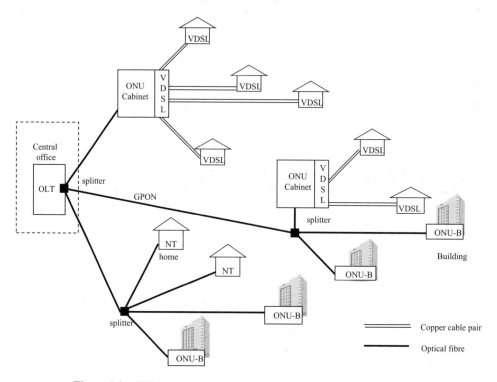

Figure 1.1 FTTCab, FTTB and FTTH fixed access network architectures.

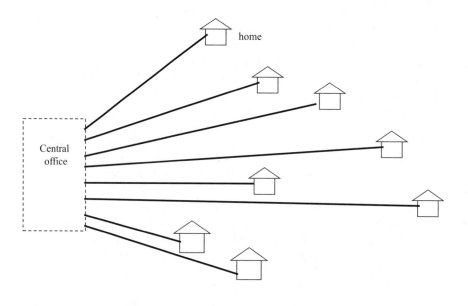

Figure 1.2 Point-to-point fibre architecture in the access network.

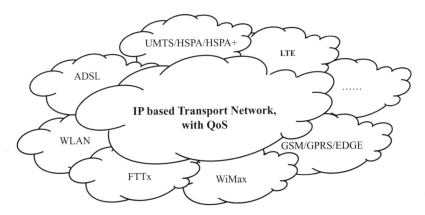

Figure 1.3 Many accesses and one core.

rate of hundreds of Mbps. Among radiomobile technologies, the global system for mobile communications (GSM) and its evolutions for data transmission, the general packet radio service (GPRS) and enhanced data rates for GSM evolution (EGDE), have been largely deployed all around the world. The third generation radiomobile system, the universal mobile telecommunication system (UMTS), with its evolutions, HSPA (high speed packet access) and HSPA+ for high bit rates data transmission, has been deployed with targeted coverage in high traffic areas, like major and minor cities. Long term evolution (LTE) is operating in many countries and is going to be deployed in others. Wireless local area networks (LANs) and WiMAX (see Section 1.3.2) are other existing technologies used mostly for data but also for voice transmission.

The different access networks, which collect all originating and terminating services, are connected to an IP-based backbone, offering a transport service with quality of service (QoS). Figure 1.3 shows the network with one core and many accesses.

1.2 Radiomobile Networks

Radiomobile networks were standardized with the aim of extending the services provided by the fixed network to mobile users, by means of a wireless terminal with the ability to move while the connection is in progress.

First generation systems, like the total access communication system (TACS), provided only the voice service, which was transmitted over the radio interface using frequency division multiple access (FDMA). The digital GSM system [1], initially standardized mainly for voice service, with its GPRS and EDGE evolutions, added new features in the access network and new nodes in the core in order to optimize data transmission.

The third generation system UMTS, standardized for multimedia, includes in its evolutions HSPA and HSPA+, which are able to reach higher bit rates and decrease latency. Finally, LTE reaches bit rates of hundreds of Mbps in downlink and lower latency times. The advanced version of LTE (LTE advanced) promises rates of Gbps.

In this section second, third and fourth generation radiomobile networks are described in terms of network architecture, access network and radio interfaces.

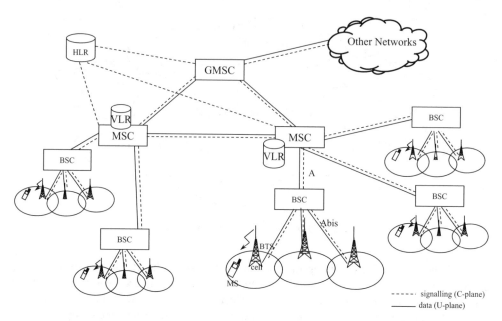

Figure 1.4   GSM network architecture.

1.2.1   GSM/GPRS/EDGE Network Architecture

Figure 1.4 shows the GSM network architecture. The first network element is the mobile station (MS), which includes the mobile terminal (MT) and the subscriber identity module (SIM). Its principal functions are transmission and reception over the radio interface, radio channels supervision, cell selection, measurements of downlink radio parameters and execution of access, authentication and handover procedures. The MS communicates through a standardized radio interface with the base transceiver station (BTS), which is the network node that realizes one or more radio coverage cells, measures uplink radio parameters, broadcasts system information and executes procedures like paging. Each BTS is connected, through the Abis interface, to the base station controller (BSC), which controls the BTS radio resources. It assigns and releases the radio channels to the mobile users, receives uplink and downlink measurements, performs intra-BSC handover, handles power control, resolves cells congestions, etc.

Because the Abis interface is not standardized, the BTSs and the connected BSCs must be from the same vendor. The BSC with its connected BTSs form a base station subsystem (BSS) and represent the GSM access network nodes. The BSSs are connected to the core network, which includes switching nodes like mobile switching centres (MSCs) and databases like the visitor location register (VLR) and home location register (HLR). BSCs are connected to the MSC through the standard A interface.

The principal functions of an MSC are: call handling, mobility handling (through interworking with VLR and HLR), paging, intra-MSC handover, inter-MSC handover, toll-ticket generation. Associated to the MSC is the VLR, which is a database containing a record for each user registered in the MSC/VLR area. Some of the MSCs are gateways (GMSC), because they are connected to the other mobile operator's networks and to the fixed network, in order to handle all the mobile–mobile, mobile–fixed and fixed–mobile calls.

The HLR is a register that stores, for each user of the mobile network, the service profile, the key for authentication and encryption, the international mobile subscriber identity (IMSI) and mobile station ISDN (MSISDN), as well as an identifier of the VLR where the user is registered. When an MS registers to the network, the VLR creates a new record with the user profile downloaded from the HLR and the MS position in terms of the location area (LA). In the HLR the identifier of the actual VLR is updated. The VLR also assigns the temporary IMSI (TMSI), which temporarily substitutes the IMSI.

The LA is a logical concept including a certain number of cells. The location area identifier (LAI) is broadcasted from the BTSs in all the cells belonging to the LA. When an MS moves from one LA to another, it performs the LA updating procedure. If the new LA belongs to a new MSC, then the new VLR downloads the user profile from the HLR and registers the new user with its LA. The HLR updates the VLR identifier and instructs the old VLR to delete the record of the user.

A GSM network with its core based on circuit-switching nodes, the MSCs, is well suited for voice but it is not for data. GSM data transmission is possible, but at a fixed bit rate of 9.6 kbps and using a voice-equivalent channel for all the duration of the call. Billing is based on the call duration and not on the amount of the exchanged data.

GPRS is the GSM evolution for data transmission. It introduces new features in the access network nodes in order to enhance data transmission speed and optimize resource allocation. In particular, GPRS needs new encoders in the BTSs and a new module, the packet control unit (PCU), in the BSC. The PCU implements radio resource management (RRM) algorithms for data transmission. GPRS also introduces new core network nodes: the serving GPRS support node (SGSN) and the gateway GPRS support node (GGSN).

The SGSN is responsible for the delivery of data packets from and to the mobile stations within its service area. Its tasks include packet routing and transfer, mobility management (attach/detach and location management), logical link management, authentication and charging functions. The location register of the SGSN stores location information and user profiles used in the packet data network of all GPRS users registered with this SGSN.

The GGSN is the node having connections with the other packet data networks. It contains routing information for the connected GPRS users. The routing information is used to tunnel packet data units (PDUs) to the MS's current point of attachment, that is the SGSN.

The BSC is connected to the SGSN through the standard Gb interface; the connection between SGSN and GGSN is the Gn interface; SGSN and GGSN are connected to the HLR through respectively the Gr and Gc interfaces; SGSN and MSC/VLR can see each other through the Gs interface. Gs and Gc interfaces are not mandatory. If Gs is present, an association between MSC/VLR and GGSN is created and it is possible to jointly handle a mobile station with packet switched and circuit switched services. Gs was introduced in order to reduce signalling over the radio interface. In fact, it is possible to carry out procedures like registration (IMSI attach) through the SGSN, combined LA and routing area (RA) updates, IMSI detach, etc. The RA is the equivalent of the LA in the GPRS domain; in general an LA contains an integer number of RAs.

A GPRS data transmission reaches a maximum download bit rate of about 50 kbps. EDGE, also called enhanced GPRS (EGPRS), is an evolution of GPRS allowing downlink bit rates up to about 240 kbps. EDGE adds new radio features to the GSM/GPRS access network nodes,

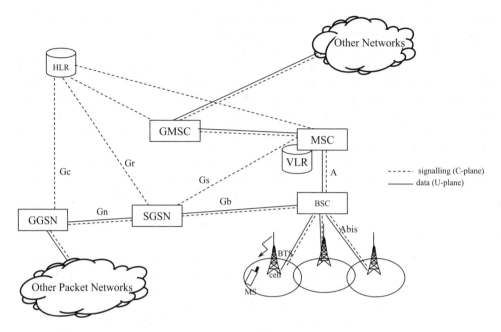

Figure 1.5   GSM/GPRS/EDGE network architecture.

and reuses the GPRS core network nodes: SGSN and GGSN. In particular, new modulators and encoders are added in the BTSs and new software in the PCU, in order to manage higher bit rate data connections.

Figure 1.5 shows the GSM/GPRS/EDGE network architecture. The access network, with its BSCs and BTSs, is shared among GSM and GPRS. The MSC-based core transports voice services and the SGSN/GGSN-based core transports data services.

1.2.2 GSM/GPRS/EDGE Access Network

The GSM/GPRS/EDGE access network, also called the GERAN (GSM EDGE radio access network), includes MSs, BTSs, BSCs, and related interfaces. The radio interface is based on frequency division duplex (FDD) and FDMA/TDMA (time division multiple access). Table 1.1 shows GSM/GPRS/EDGE working frequencies in different countries of the world.

Table 1.1   GSM working frequencies in different countries of the world

Band	Uplink (MHz)	Downlink (MHz)
GSM 900	880–915	925–960
GSM 1800	1710–1785	1805–1880
PCS (personal communication service) 1900	1850–1910	1930–1990
Cellular 850	824–849	869–894

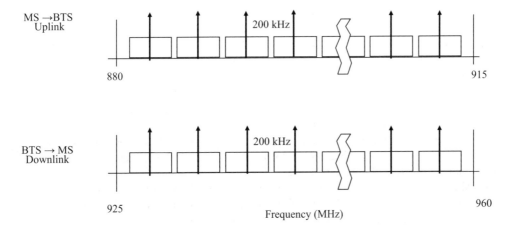

Figure 1.6   Division of the GSM 900 band into 200 kHz carriers.

In Europe, Africa, the Middle East and Asia most of the providers use 900 MHz and 1800 MHz bands. In North America, GSM operates on the bands of 850 MHz and 1900 MHz. GSM at 850 and 1900 MHz is also used in many countries of South and Central America.

All over the world, a refarming process of the radiomobile spectrum is going on, which is a rearrangement of the frequencies used for mobile services. For example, the 900 MHz band used for GSM is now available also for third generation (UMTS) services. FDMA in GSM contemplates the division of the assigned spectrum into carriers spaced 200 MHz apart. Figure 1.6 shows the division of the GSM 900 band into 200 kHz carriers.

The theory of GSM frequency planning introduces the concept of cluster, which is a group of cells using all the available carriers. Cellular coverage is based on the repetition of the cluster. Figure 1.7 shows an example with the cluster and the relative theoretical frequency planning.

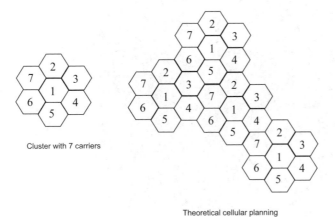

Figure 1.7   Example of theoretical frequency planning with repetition of the cluster.

As very often happens, the reality is far from the theory. The goal of cell planning is to guarantee the availability of radio resources that satisfy QoS for the provision of a set of services in an area target. In general, forms and dimensions of each cell are different and in each cell one or more carriers are switched on, depending on the requirements of coverage, capacity and performances. When GSM 900 and GSM 1800 coexist, an overlay/underlay technique is used for coverage. The underlay coverage is in general at 900 MHz and covers a higher area than overlay cells working at 1800 MHz. Overlay and underlay cells share sites, antenna systems and control channels.

Among cells of different dimensions, there are the macro, micro and pico cells depending on the size. Micro and pico cells are used to solve traffic peaks in small areas. Figure 1.8 shows an example of macro and micro coverage.

GSM multiple access is FDMA/TDMA and is based on a frame structure of eight time slots per carrier, as shown in Figure 1.9. The frame duration is 4.6 ms; the slot duration is 577 μs; the signal burst is contained in one time slot and lasts 546 μs.

The transmission and reception frames are shifted by three time slots. This allows the mobile station transceiver to transmit over the uplink, move to the downlink frequency, receive the downlink signal and make measurements over the other radio channels. This process is shown in Figure 1.10.

The GSM frames are grouped together to form multiframes, superframes and iperframes. This temporal structure allows the establishment of a time schedule for operation and network synchronization. In particular, one multiframe can be formed of 26 or 51 frames; one superframe lasts 6.12 s and is formed of 51 multiframes of 26 frames or 26 multiframes

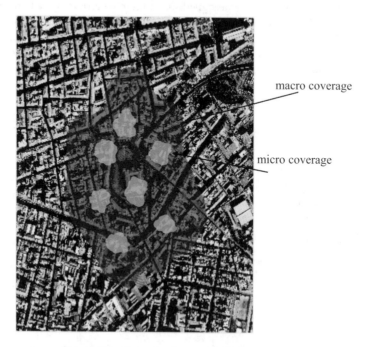

Figure 1.8   Example of macro and micro coverage.

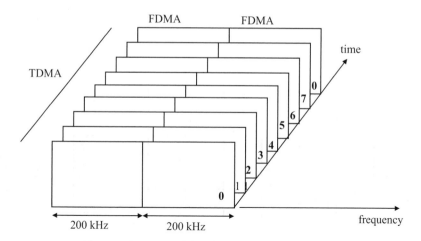

Figure 1.9   FDMA/TDMA multiple access in GSM.

of 51 frames; one iperframe is formed of 2048 superframes and lasts 3 h, 28 min, 53 s and 760 ms.

GSM access includes, as an optional feature, frequency hopping (FH). The goal of frequency hopping in GSM is to obtain an intrinsic diversity in frequency that protects the transmission from effects like rapid fluctuations of the radio channel or cochannel interferences. There are a total of 63 different hopping algorithms available in GSM.

When the BTS orders the MS to switch to the frequency hopping mode, it also assigns a list of channels and the hopping sequence number (HSN), which corresponds to the particular hopping algorithm that will be used. Figure 1.11 shows the principle of frequency hopping.

1.2.2.1   GSM Physical and Logical Channels

A physical channel in the GSM access network is identified from the time slot in the frame, the frame number, an FH sequence (if FH is active). A logical channel is dedicated to the

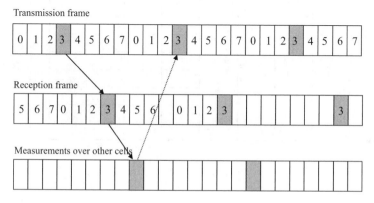

Figure 1.10   GSM transmission and reception.

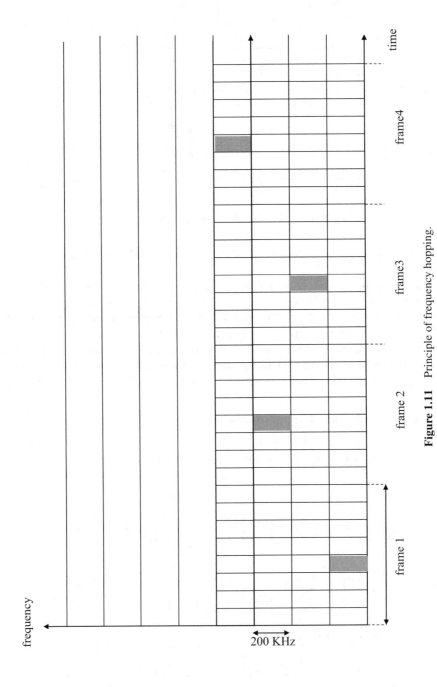

Figure 1.11 Principle of frequency hopping.

transmission of specific information, using a mapping over appropriate physical resources. Logical channels are divided into traffic channels (TCHs), carrying voice and data, and control channels (CCHs), carrying control information.

GSM TCHs are:

- TCH/FS: full rate (FR) speech
- TCH/HS: half rate (HR) speech
- TCH/F9.6: data at 9.6 kbps (FR)
- TCH/F4.8: data at 4.8 kbps (FR)
- TCH/F2.4: data at 2.4 kbps (FR)
- TCH/F1.2: data at 1.2 kbps (FR)

Also half rate data channels at 4.8 and 2.4 kbps are defined.

Full rate traffic channels use one time slot per frame; half rate channels use one time slot each two frames, occupying half capacity. The gross bit rate of a full rate channel is 22.8 kbps; the gross bit rate of an half rate channel is 11.4 kbps.

CCHs are divided into broadcast channels (BCHs), carrying broadcast information, common control channels (CCCHs), carrying common signalling information, and dedicated control channels (DCCHs) carrying signalling information dedicated to a user.

Figure 1.12 shows control and traffic channel mapping.

In downlink, BCHs are:

- FCCH: frequency correction channel, carrying a frequency reference signal
- SCH: synchronization channel, carrying a frame synchronization reference signal and a base station identity code (BSIC)

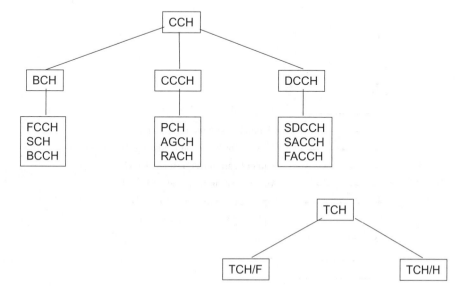

Figure 1.12   GSM logical channels mapping.

- BCCH: broadcast control channel, carrying a cell global identity (CGI), location area identity (LAI), frequency hopping algorithm, references for control channels of adjacent cells and other cell parameters

CCCHs are:

- PCH: paging channel, where the downlink is sent to search for an MS having an incoming call and paging is broadcast over all the cells belonging to the MS location area
- AGCH: access grant channel, where the downlink is used to allocate a standalone dedicated control channel (SDCCH) to the MS
- RACH: random access channel, where the uplink carries user access information and is used from the MS to request an SDCCH allocation

DCCHs are:

- SDCCH: standalone dedicated control channel, which is bidirectional and uses to execute signalling procedure like location area updating, TMSI allocation, attach and detach and occupies an eighth of one slot
- SACCH: standalone associated control channel, which is bidirectional and carries control information related to an active connection like measurement reports in uplink and power control and timing advance in downlink
- FACCH: fast associated control channel, which is bidirectional and uses one TCH by substituting signalling to the traffic (frame stealing), in general carrying handover information

Figures 1.13, 1.14 and 1.15 show examples of logical channels usage over the radio interface in location area updating, mobile-terminated and mobile-originating call procedures.

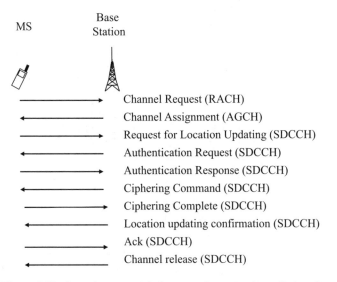

Figure 1.13   Location area updating procedure over the radio interface.

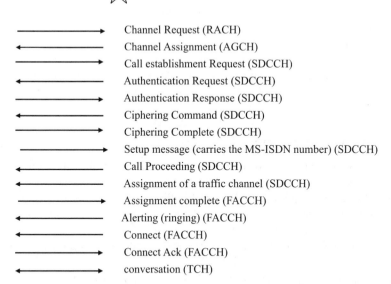

MS

Base
Station

Channel Request (RACH)
Channel Assignment (AGCH)
Call establishment Request (SDCCH)
Authentication Request (SDCCH)
Authentication Response (SDCCH)
Ciphering Command (SDCCH)
Ciphering Complete (SDCCH)
Setup message (carries the MS-ISDN number) (SDCCH)
Call Proceeding (SDCCH)
Assignment of a traffic channel (SDCCH)
Assignment complete (FACCH)
Alerting (ringing) (FACCH)
Connect (FACCH)
Connect Ack (FACCH)
conversation (TCH)

Figure 1.14   Mobile-originating call procedure over the radio interface.

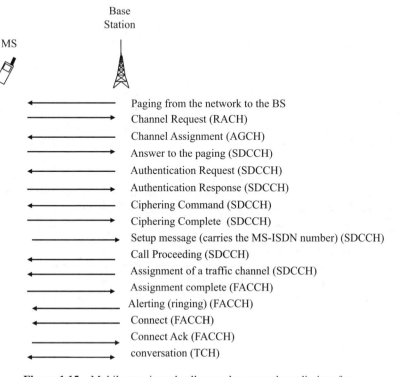

MS

Base
Station

Paging from the network to the BS
Channel Request (RACH)
Channel Assignment (AGCH)
Answer to the paging (SDCCH)
Authentication Request (SDCCH)
Authentication Response (SDCCH)
Ciphering Command (SDCCH)
Ciphering Complete (SDCCH)
Setup message (carries the MS-ISDN number) (SDCCH)
Call Proceeding (SDCCH)
Assignment of a traffic channel (SDCCH)
Assignment complete (FACCH)
Alerting (ringing) (FACCH)
Connect (FACCH)
Connect Ack (FACCH)
conversation (TCH)

Figure 1.15   Mobile-terminated call procedure over the radio interface.

1.2.2.2 GSM/GPRS Modulation

The GSM and GPRS modulation technique is a Gaussian minimum shift keying (GMSK); it is a minimum shift keying (MSK) with a premodulation Gaussian filter.

The GMSK was chosen as a compromise between spectral efficiency, realization complexity and low emission of spurious radiations (with low adjacent channel interference). The GMSK improves the spectral efficiency respect to the MSK because the power spectral density (PSD) presents a reduced main lobe compared to the MSK. The modulation rate is $270 \times (5/6)$ kbauds.

1.2.2.3 GPRS Radio Interface

GPRS is the GSM evolution for data transmission [2]. It implies the introduction of new network nodes in the core network and new features in the access network to increase the maximum bit rate and optimize the usage of radio resources for data transmission. Physical channels are not allocated permanently to a GPRS connection, but only when data have to be transmitted over the radio interface.

GPRS shares with GSM the radio resources and introduces the following possibilities:

- More than one time slot of a TDMA frame can be assigned to an MS.
- More than one MS can be multiplexed on the same time slot.
- Radio resources are separately assigned to uplink and downlink (asymmetric transmission and reception).
- GSM and GPRS can use the same time slot at different times.
- GSM has higher priority than GPRS.
- GPRS assigned resources can be dropped.

In GPRS, four coding schemes (CSs) with different coding rates are introduced. The higher the coding rate, the higher is the net bit rate per time slot. On the other hand, with higher coding rates a higher signal to noise ratio (SNR) at the receiver side is required to achieve the same bit error rate (BER). Table 1.2 shows the coding schemes with the corresponding net bit rate per time slot at the radio link control (RLC) layer.

A full duplex MS could use up to eight slots per frame, but most GPRS terminals are half duplex. A half duplex terminal is not able to transmit and receive at the same time. The maximum number of time slots that a terminal can use in a frame is five.

Table 1.2 GPRS coding schemes

Coding scheme (CS)	Code rate	RLC bit rate per time slot (kbps)
CS1	1/2	8
CS2	~2/3	12
CS3	~3/4	14.4
CS4	1	20

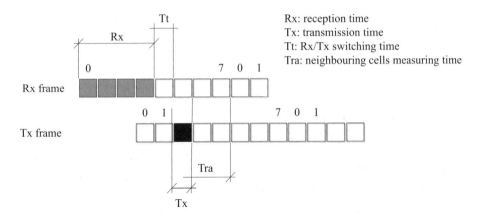

Figure 1.16   Example of a GPRS asymmetric transmission (4 + 1).

Figure 1.16 shows the usage of four slots in downlink and one slot in uplink (4 + 1) from a simplex terminal. The figure shows the reception time in four slots in downlink (Rx), the time needed to switch to the transmission frame (Tt), the transmission time in one slot (Tx) in uplink, the neighbouring cells measuring time (Tra) and the lasting two slots.

The standard defines GPRS logical channels, also called packed data logical channels (PDCH).

They are:

- Packet common control channels (PCCCHs):
  - Packet random access channel (PRACH): for random access (uplink)
  - Packet paging channel (PPCH): for paging (downlink)
  - Packet access grant channel (PAGCH): for access grant (downlink)
  - Packet notification channel (PNCH): for point-to-multipoint-multicast (PTM-M) notification (downlink)
- Packet broadcast control channel (PBCCH): used to broadcast system information to GPRS mobile stations (downlink)
- Packet traffic channels (PTCHs):
  - Packet dedicated traffic channel (PDTCH): bidirectional, carries packet data traffic
- Packet dedicated control channels (PDCCHs):
  - Packet associated control channel (PACCH): bidirectional, carries associated control to a data connection
  - Packet timing advance control channel (PTCCH): bidirectional, carries in uplink access bursts that allow the network to calculate the timing advance, which is then transmitted over the same channel in a downlink timing advance value corresponding to the time a signal takes from the mobile station to the BTS

An operator can choose not to reserve dedicated resources to GPRS control channels. In this case, packet control channels are not configured over the radio interface and GPRS service shares control channels with GSM: a GPRS mobile station receives specific GPRS system information on the broadcast control channel (BCCH).

In the case where packet control channels are configured, a GPRS mobile station monitors the PBCCH, where specific GPRS system information is sent other than some information related to circuit switched services. If that happens, a GPRS mobile is not requested to monitor the BCCH.

1.2.2.4 EDGE Radio Interface

EDGE is an evolution of GPRS, especially in relation to the radio interface. It is also called enhanced GPRS (EGPRS). The principal goal is to increase the data rate through improved spectral efficiency.

The main features introduced in EDGE are:

- Eight phase shift keying (8PSK) modulators other than GMSK modulators,
- New modulation and coding schemes (MCSs),
- Link adaptation: varies the modulation and coding scheme (MCS) in relation to the quality of the radio channel, and
- Hybrid automatic repeat request (HARQ): combines two techniques, FEC (forward correcting coding) with ARQ (automatic repeat request).

Table 1.3 shows the nine EDGE modulation and coding schemes (MCSs) with the related RLC data rates per time slot. The higher the data rate per time slot, the higher is the required SNR at the receiver side.

The maximum downlink bit rate per user is about 240 kbps, obtained considering four assigned time slots encoded with MCS9. Each MCS belongs to a class (A, B, C). The link adaptation varies if the channel conditions degrade and the modulation and coding scheme is within the same class for the retransmission of the same packet. Figure 1.17 shows a qualitative example of link adaptation. In the example, the data transmission starts with MCS6. If the received SNR decreases, the block error rate (BLER) increases, and the modulation and coding scheme is switched to MCS3, at the expense of the throughput.

Table 1.3 EDGE modulation and coding schemes

MCS	Family	Modulation	Code rate	RLC data rate per time slot (kbps)
MCS9	A	8PSK (3 bits/	1/2	59.2
MCS8	A	modulation	~2/3	54.4
MCS7	B	symbol)	~3/4	44.8
MCS6	A		1	29.6
MCS5	B		1/2	22.4
MCS4	C	GMSK (1 bit/	~2/3	17.6
MCS3	A	modulation	~3/4	14.8
MCS2	B	symbol)	1	11.2
MCS1	C		1	8.8

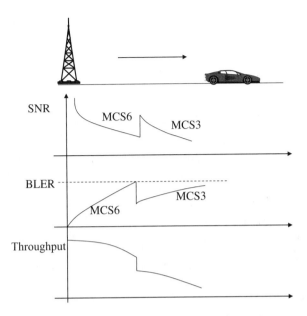

Figure 1.17   Qualitative example of link adaptation.

EDGE network architecture is the same as GPRS, with SGSN and GGSN nodes for data transport and the same access network nodes with some added features, like the new 8PSK modulator and encoders in the BTSs and new protocols in the BSC, which control EDGE transmissions.

1.2.3   UMTS/HSPA/HSPA+ Network Architecture

In 1995, the European Telecommunication Standard Institute (ETSI) started to work on the UMTS third generation system [3,4]. In 1997, UMTS was defined in [5] as follows:

> UMTS will be a mobile communications system that can offer significant user benefits including high-quality wireless multimedia services to a convergent network of fixed, cellular and satellite components. It will deliver information directly to users and provide them with access to new and innovative services and applications. It will offer mobile personalized communications to the mass market regardless of location, network and terminal used.

Contrary to the GSM, which was built for voice, UMTS was built for 'multimedia services' and 'personalized communications'.

In January 1998, the choice of the radio interface was frozen:

- Wideband CDMA (code division multiple access) in paired bands (FDD mode)
- TD-CDMA (time division-CDMA) in unpaired bands (TDD mode)

In December 1998, the standardization bodies from Europe, Japan, Korea and the United States created the 3GPP (3rd Generation Partnership Project), which now counts six organizational

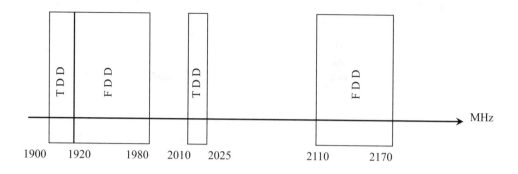

FDD frequency Division Duplex
TDD Time Division Duplex

Figure 1.18   European bands designed for UMTS.

partners (the Japanese ARIB, the American ATIS, the Chinese CCSA, the European ETSI, the Korean TTA and the Japanese TCC) and many market representation partners. The 3GPP original mandate was to develop specifications for the UMTS third generation mobile system based on a new access network and a core network evolved from the GSM/GPRS/EDGE core network.

In the World Radiocommunication Conference held in Geneva in 1997, the bands 1885–2025 MHz and 2110–2200 MHz were identified for 3G systems. In Europe, 215 MHz have been assigned to UMTS, as shown in Figure 1.18. In Europe, only the UMTS FDD band is used.

Because the initially defined bands were already used in various regions of the world, some other bands for 3G systems were added to meet the needs of various countries. The band around 2100 MHz is used in Europe, China, Korea, Japan, Australia, India and Latin America; in North America UMTS works at 1900 and 850 MHz; in Japan other than 2100 MHz also bands at 1700 and 800 MHz are used. Refarming, which provides a rearrangement of the frequencies assigned to mobile services, opens to a possible usage also of the actual GSM bands for UMTS.

Figure 1.19 shows 3GPP releases from R99 to R11 with their principal characteristics. It shows that 3GPP standardized UMTS, its evolution HSPA/HSPA+, LTE and LTE advanced.

Figure 1.20 shows the UMTS network architecture standardized in release 99 (R99) of the 3GPP standard. The UMTS network can be divided into an access network, called UMTS terrestrial radio access network (UTRAN), and a core network. UMTS R99 introduces a radio access network based on code division multiple access (CDMA) radio technology.

The mobile station (MS) communicates through a standardized radio interface with the node B, which is a node equivalent to the BTS in GSM. Node B handles more than one cell, makes measurements over the uplink, broadcasts cell parameters and executes procedures like paging and inner-loop power control. Each node B is connected, through the Iub interface, to

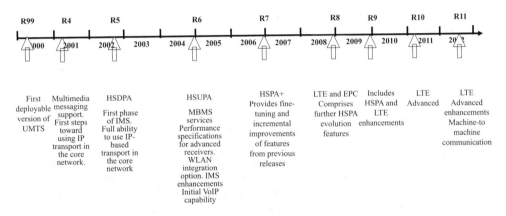

Figure 1.19 3GPP releases.

the radio network controller (RNC), which assigns and releases radio channels to the mobile users, receives uplink and downlink measurements, performs the handover procedure, handles outer-loop power control, etc. The Iub interface is not standard and therefore the node Bs with the connected RNCs must be from the same vendor. The RNC with its connected node Bs form a radio network system (RNS).

The RNSs are connected to the core network through the standard Iu interface, which is divided into two branches: Iu-CS for circuit switched (CS) services and Iu-PS for packet

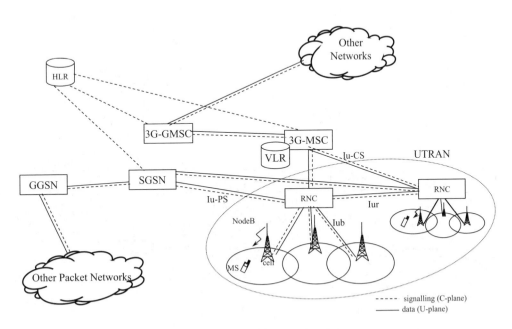

Figure 1.20 R99 UMTS network architecture.

switched (PS) services. The R99 core network is based on a moderate evolution of the GSM/GPRS/EDGE core.

Voice and data traffic, as well as all the radio signalling protocols, are carried transparently from the MS to the RNC. Separation of packet switched and circuit switched services occurs only at the Iu interface. Iu-CS connects circuit switched services (i.e., voice and video calls) to the CS domain based on 3G-MSCs; packet switched services (data) are carried through SGSN and GGSN nodes.

An important evolution of the circuit switched core network was introduced from 3GPP in release 4. Release 4 CS separation of control and transport functions is a first evolution towards an IP-based core network. In this architecture, a connectivity layer is introduced for user data transport. The key element in this layer is the media gateway (MGW). It is connected to a generic packet switching core and opens to any type of underlying transport network. Nb is the interface connecting the MGWs, and supports ATM (asynchronous transfer mode) or IP transport. MSC servers control the MGWs through the Mc interface. The interface between MSC servers is Nc, and supports call control over IP or ATM. When a call is originated in the UMTS access network, the MSC servers establish a connection between the originating MGW and the MGW that is closest to the end-destination. Through this connection, which can be physical or logical depending on the underlying transport network, real time traffic is transported. Figure 1.21 shows the UMTS network architecture provided in release 4 of the 3GPP standard.

Release 5 includes the introduction of the IP multimedia subsystem (IMS), the architecture for IP multimedia services. IMS uses IP to transport signalling and user data and is based on the session initiation protocol (SIP).

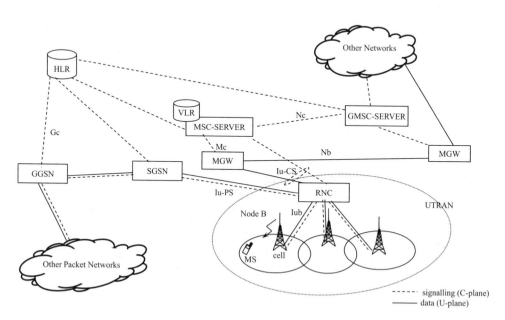

Figure 1.21　UMTS network architecture in release 4.

1.2.4   UMTS/HSPA/HSPA+ Access Network

The UMTS/HSPA/HSPA+ access network includes mobile stations (MSs), node Bs, radio network controllers (RNCs) and related interfaces. The radio interface is based on code division multiple access (CDMA), in both FDD and TDD modes.

In the FDD mode, one band is assigned to the uplink and one different band is assigned to the downlink. Each bandwidth is 5 MHz wide. In the TDD mode, a 5 MHz band is shared in time between the uplink and downlink. In most countries only UMTS FDD was installed.

1.2.4.1   CDMA Basics

CDMA is a multiple access technique based on the assignment of different codes to different users. A simplified transmission scheme based on CDMA is illustrated in Figure 1.22.

The signal at bit rate $R_b = 1/T_b$ enters in the *channel coding* block with coding rate k/n, with $k/n \leq 1$. This means that for every k bits of useful information, the encoder generates n bits of data. The bit rate of the signal at the output of the channel coding block is $R'_b = 1/T'_b = (n/k) R_b$.

The key element is the spreading module, which performs spread spectrum with the spreading factor (SF). The spreading module works as follows. Each input bit is multiplied by a

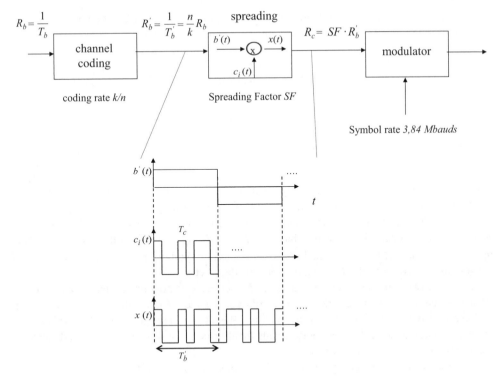

Figure 1.22   Transmission scheme based on CDMA.

codeword (spreading code) having the duration of the bit but which is composed of a number of chips equal to the SF. A chip is one of the pulses forming the codeword $c(t)$.

Multiple access is achieved by assigning different spreading codes to different users. If the codes are orthogonal and the transmission among the different users is synchronous, then each signal can be perfectly decoded. Two codes $c_i(t)$ and $c_j(t)$ are orthogonal if:

$$\frac{1}{T_b'} \int_{T_b'} c_i(t) \cdot c_j(t) \mathrm{d}t = E\left[c_i(t) \cdot c_j(t)\right]\big|_{T_b'} = \begin{cases} 1 & \text{if} \quad i = j \\ 0 & \text{if} \quad i \neq j \end{cases} \tag{1.1}$$

The signal encoded with spreading code $c_i(t)$ is:

$$x(t) = b'(t) \cdot c_i(t) \tag{1.2}$$

If N synchronous encoded signals having the same rate R_b' are transmitted and neither noise nor attenuation is taken into account, then the signal at the receiver side is the sum of the N signals:

$$z(t) = \sum_{k=1}^{N} x_k(t) = \sum_{k=1}^{N} b_k'(t) \cdot c_k(t) \tag{1.3}$$

To decode the bit b_i' transmitted from the ith user during the interval T_b', the received signal $z(t)$ is multiplied by $c_i(t)$ and then the obtained signal is averaged in T_b':

$$E\left[z(t) \cdot c_i(t)\right]\big|_{T_b'} = \frac{1}{T_b'} \int_{T_b'} \left(\sum_{k=1}^{N} x_k(t)\right) c_i(t)\, \mathrm{d}t = \frac{1}{T_b'} \int_{T_b'} \left(\sum_{k=1}^{N} b_k' c_k(t)\right) c_i(t)\, \mathrm{d}t$$

$$= \sum_{k=1}^{N} b_k' \frac{1}{T_b'} \int_{T_b'} c_k(t) c_i(t)\, \mathrm{d}t = b_i' \tag{1.4}$$

Equation (1.4) shows that with orthogonal codes and synchronous transmissions, the transmitted signals can be perfectly decoupled. If pseudo-orthogonal codes are used, or the transmitted signals are not synchronous, then an interference component must be taken into account. The encoded signal, which has rate $R_c = SF \cdot R_b'$, goes into the modulator. The modulator clock is 3.84 MHz (3.84×10^6 modulation symbols per second), obtained by considering a carrier of 5 MHz and a 0.3 roll-off filter factor.

In UMTS, two families of codes are used: the perfectly orthogonal channellization codes (Walsh–Hadamard codes) and the pseudo-orthogonal codes. Channellization codes have a variable spreading factor (SF) according to R_b' and the rate k/n. They are also called orthogonal variable spreading factor (OVSF) codes, channellization codes or spreading codes, and increase the amplitude of the transmission band.

Figure 1.23 shows the construction of OVSF codes. The length of OVSF codes is always a power of two. Pseudo-orthogonal codes are scrambling codes and do not change the transmission bandwidth. In downlink, scrambling codes are associated with different cells; spreading

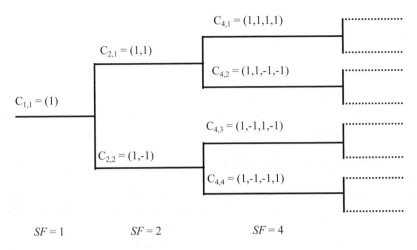

Figure 1.23 Construction of OVSF codes.

codes are associated with different users within the same cell. The number of scrambling codes for the downlink is limited to 512 Gold codes, which are used for cell planning. In uplink, scrambling codes discriminate different users; spreading codes discriminate different communications of the same user. The families of scrambling codes used for uplink are made by millions of different codes and is not necessarily code planning.

1.2.4.2 UMTS Logical, Transport and Physical Channels

Logical channels are the resources offered from the medium access control (MAC) layer to the upper layers, and are characterized by the nature of their information content; they carry information related to both user and control planes. Logical channels in UMTS are:

- Broadcast control channel (BCCH): carries system information and network configuration parameters
- Common control channel (CCCH): carries bidirectional control information for MSs not in the connected mode
- Paging control channel (PCCH): used to page an MS
- Dedicated control channel (DCCH): transfers information in the control plane for terminals in the connected mode
- Dedicated traffic channel (DTCH): transfers information in the user plane for terminals in the connected mode

Transport channels are resources offered from the physical to the MAC layer. Transport channels in UMTS can be shared or dedicated. Transport channels in UMTS are:

- Dedicated channel (DCH): downlink and uplink, carries user plane or control plane information
- Broadcast channel (BCH): downlink, sends cell and system information

- Forward access channel (FACH): downlink, used for signalling and data
- Paging channel (PCH): downlink, carries paging
- Random access channel (RACH): uplink, carries control information between the MS and eNB (evolved node B); is subject to collisions

Physical channels are the resources used for transmission over the radio interface. They are defined by a carrier frequency, a forward error correction (FEC) code, a scrambling code and a spreading code. Physical channels are:

- Physical random access channel (PRACH): uplink, carries random access preambles during the random access procedure
- Dedicated physical channel (DPCH): uplink and downlink, carries signalling (dedicated physical control channel, or DPCCH) and data (dedicated physical data channel, or DPDCH) dedicated to a connection
- Primary common control physical channel (PCCPCH): downlink, carries BCH
- Secondary common control physical channel (SCCPCH): downlink, carries PCH and FACH

Figure 1.24 shows downlink and uplink channels at logical, transport and physical layers and their mapping. In the figure are also shown the following channels:

- Synchronization channel (SCH): downlink, to synchronize the mobile stations to the network
- Indicator channels (ICHs): signalling entities with Boolean value

Examples of indicators are:

- Acquisition indicator channel (AICH): downlink, for the response to the preamble in the PRACH
- Paging indicator channel (PICH): downlink, indicates to the MS in the sleep mode to listen to the paging channel in the subsequent frame; the sleep mode is a discontinuous reception mode to save batteries

When an MS switches on, it searches and selects one cell through the synchronization signals, then derives system information from the PCCPCH, executes the random access procedure and sends signalling to register to the network.

Figure 1.25 shows an example with the use of logical, transport and physical channels in the radio resource control (RRC) connection setup procedure. The RRC protocol performs allocation and release of the radio resources, admission and congestion control, etc. It is at layer 3 in the control plane protocol stack.

1.2.4.3   UMTS Modulations

The FDD version of UMTS uses quadrature phase shift keying (QPSK) modulation in downlink and dual binary phase shift keying (BPSK) modulation in uplink. Dedicated physical channels DPDCH and DPCCH are multiplexed in time in downlink, and each constellation symbol carries DPDCH or DPCCH, transmitted with the same SF (from 4 to 512). In uplink, the

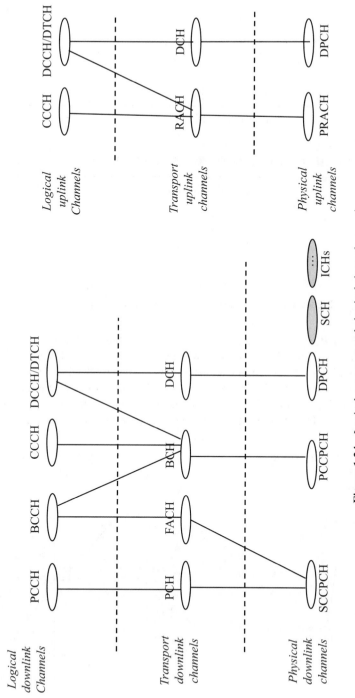

Figure 1.24 Logical, transport and physical channels mapping.

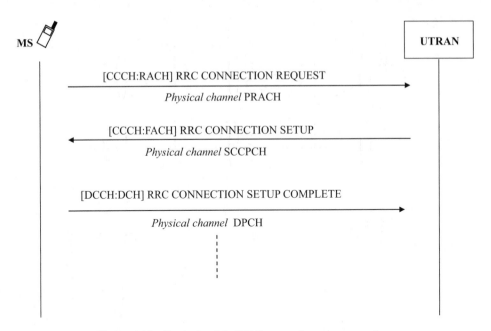

Figure 1.25   Example of the RRC connection setup procedure.

I-branch carries DPCCH (with fixed SF = 256) and the Q-branch carries DPDCH (with a variable SF from 4 to 256). Figure 1.26 shows the modulations in the UMTS FDD mode.

In uplink, to maintain the synchronization between MS and BTS, DPCCH is transmitted also during DPDCH pauses. The physical resources are allocated in UMTS for an interval called the transmission time interval (TTI). The shortest TTI is 10 ms, but it can also be of 20 ms, 40 ms and 80 ms. In UMTS, information is structured in frames of 10 ms and 15 time slots.

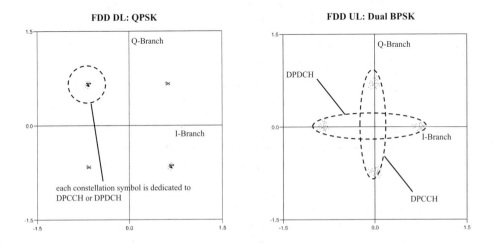

Figure 1.26   Modulations in the UMTS FDD mode.

1.2.4.4   HSPA

High speed packet access (HSPA) is the UMTS evolution for high bit rate data transmission. It can be divided into its downlink and uplink versions: high speed downlink packet access (HSDPA) and high speed uplink packet access (HSUPA).

HSDPA has been introduced in release 5 of the 3GPP standard and includes:

- Addition of 16-QAM (quadrature amplitude modulation) to the QPSK;
- Adaptive modulation and coding (AMC), with peak bit rate up to 14.4 Mbit/s;
- Fixed spreading factor SF = 16 and multicode transmission;
- Introduction of new radio channels; and
- Some RNC functionalities are moved to node B.

HSDPA introduces a new transport channel, the high speed downlink shared channel (HS-DSCH), where the TTI is reduced at 2 ms. Different users can be multiplexed in adjacent TTIs in the same HS-DSCH. The HS-DSCH is mapped at the physical layer into the high speed physical downlink shared channel (HS-PDSCH).

In the uplink, an acknowledgement of the radio block (ACK/NACK) is carried in a physical dedicated control channel, the high speed dedicated physical control channel (HS-DPCCH). The HS-DPCCH also carries the channel quality indicator (CQI), corresponding to the modulation and coding scheme (MCS) and transport block size (TBS), for which the estimated received downlink transport block error rate (BLER) shall not exceed 10%. TBS is the amount of data carried in a TTI.

Figure 1.27 shows that different users are multiplexed in the HS-DSCH with 2 ms TTIs. In the uplink, the HS-DPCCH carries ACK/NACKs and CQIs.

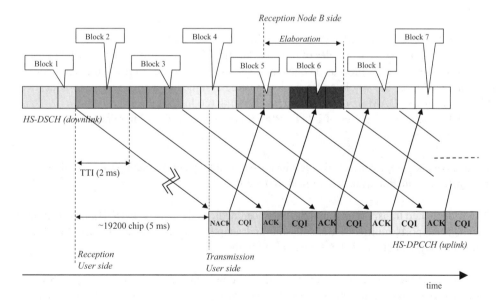

Figure 1.27   HSDPA transmission on the HS-DSCH and feedbacks.

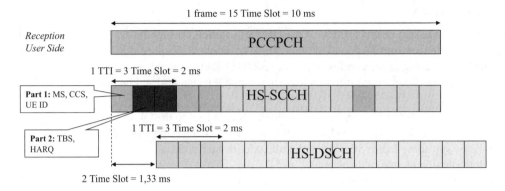

Figure 1.28   Scheduling information in HSDPA.

Scheduling information is carried in a new downlink shared control channel, the high speed shared control channel (HS-SCCH). It uses a code with SF = 128 and sends, in an interval of 2 ms, the transport format (TF), which represents the information needed to demodulate and decode (modulation and coding scheme, transport block size) the data sent to a mobile station on the HS-DSCH. In particular, scheduling information is divided into two parts:

- Part 1: lasts for one time slot and carries the information necessary to demodulate the HS-DSCH. It contains the mobile station MAC identifier, the modulation and the channellization code Set (CCS), which identifies the assigned spreading codes.
- Part 2: is superimposed on to the HS-DSCH and carries the information necessary to decode the HS-DSCH, like the transport block size (TBS) and the hybrid automatic repeat request (HARQ) scheme.

Figure 1.28 shows the HS-SCCH carrying scheduling information.

Table 1.4 shows some examples of transport formats. The first row shows the R99 transport format for the 384 kbps bit rate. The last row shows that, in order to reach the maximum bit rate of 14.4 Mbps, the whole cell capacity must be assigned to a single user (15 codes with SF = 16) with 16 QAM modulation and no encoding (coding rate = 1).

HSUPA uses most features of UMTS R99. It has been introduced in release 6 of the 3GPP standard and provides:

- Adaptive encoding, with multicode transmission and a peak bit rate up to 5.76 Mbps
- Some RNC functionalities are moved to node B

A new dedicated transport channel, the enhanced dedicated channel (E-DCH), has been introduced. It supports multicode transmission and adaptive encoding. To support HSUPA, the following physical channels have been added to the radio interface:

- Enhanced HARQ indicator channel (E-HICH): downlink, used to send HARQ ACKs
- Enhanced relative grant channel (E-RGCH): provides relative step up/down scheduling commands

Table 1.4 HSDPA transport formats

Modulation	TBS (bit)	TTI (ms)	Coding rate	RLC data rate (Mbps)	Number of codes
QPSK (R99)	3840	10	1/3	0.384	1 (SF = 8)
QPSK	317	2	1/3	0.16	1 (SF = 16)
QPSK	461	2	1/2	0.23	1 (SF = 16)
QPSK	931	2	1/2	0.46	2 (SF = 16)
QPSK	1483	2	1/2	0.74	4 (SF = 16)
QPSK	2279	2	1/2	1.14	5 (SF = 16)
QPSK	3319	2	~0.7	1.65	5 (SF = 16)
16-QAM	3565	2	~0.4	1.8	5 (SF = 16)
16-QAM	4664	2	1/2	2.3	5 (SF = 16)
16-QAM	7168	2	3/4	3.6	5 (SF = 16)
16-QAM	11 418	2	3/4	5.7	8 (SF = 16)
16-QAM	14 411	2	3/4	7.2	10 (SF = 16)
16-QAM	17 237	2	3/4	8.6	12 (SF = 16)
16-QAM	21 754	2	3/4	10.9	15 (SF = 16)
16-QAM	25 558	2	~0.9	12.8	15 (SF = 16)
16-QAM	28 776	2	1	14.4	15 (SF = 16)

- Enhanced absolute grant channel (E-AGCH): provides absolute scheduling for the user equipment (UE)
- Enhanced dedicated physical data channel (E-DPDCH): carries user plane information
- Enhanced dedicated physical control channel (E-DPCCH): carries control plane information

Figure 1.29 shows HSDPA and HSUPA logical, transport and physical channels and their mapping.

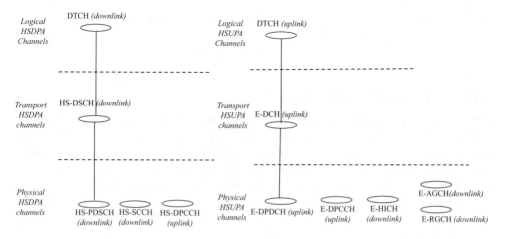

Figure 1.29 HSDPA and HSUPA logical, transport and physical channels and their mapping.

Table 1.5 HSPA and HSPA+ principal features

3GPP release	DL		UL	
	Features	RLC bit rate (Mbps)	Features	RLC bit rate (Mbps)
R6	16-QAM	14.4	Dual BPSK	5.76
R7	64-QAM	21.1	16-QAM	11.5
	16-QAM, 2 × 2 MIMO	28.8	16-QAM	11.5
R8	64-QAM, 2 × 2 MIMO	42.2	16-QAM	11.5
R9	64-QAM, 2 × 2 MIMO, dual carrier (10 MHz)	84	16-QAM, dual carrier	23
R10	64-QAM, 2 × 2 MIMO, four carrier aggregation (20 MHz)	168	16-QAM, dual carrier	23
R11	64-QAM, 4 × 4 MIMO, four carrier aggregation (20 MHz)	336	64-QAM, dual carrier, 2 × 2 MIMO	69

1.2.4.5 HSPA+

The evolved high speed packet access is also called HSPA+ and has been standardized starting from release 7. It introduces new features at the physical layer to increase the maximum uplink (UL) and downlink (DL) bit rates and new features at the MAC layer to enhance packet connectivity performances.

The principal features introduced at the physical layer up to release 11 are:

- Higher order QAMs: used to increase uplink and downlink bit rates.
- Multiple input multiple output (MIMO) antennas: used to enhance the received signal to noise ratio (SNR) or to multiply the bit rate. An introduction to MIMO is given in Section 1.2.6.4 of this book.
- Carrier aggregation: more carriers of 5 MHz each are aggregated to multiply the bit rate.

Table 1.5 shows the features introduced in HSPA+ up to 3GPP release 11.

1.2.5 LTE Network Architecture

LTE is a new system standardized by 3GPP. The work on LTE started with a workshop in Toronto, on 2 and 3 November 2004. During the workshop, manufacturers, researchers and operators contributed to identify the high level requirements for the evolution of UTRAN. The focus was mainly on the radio access with several proposals for radio evolution. From December 2004 to June 2006 feasibility studies on the LTE system were conducted.

LTE requirements, identified from the beginning, concerned services (support of voice and multimedia over IP, high uplink and downlink bit rates, low latency), radio performances (scalable bandwidth, usage of MIMO to improve the throughput), costs (simplified network architecture, transport of user plane and control plane over IP) and interworking with existing radiomobile networks [6].

The detailed standard work started in June 2007. The 3GPP goal was 'to develop a framework for the evolution of the 3GPP radio-access technology towards a high-data-rate, low-latency and packet-optimized radio-access technology'.

In parallel to the project for the definition of LTE radio access, a 3GPP project related to the core network started. The project was called system architecture evolution (SAE) with the aim of standardizing the evolved packet core (EPC). EPC is an all IP network which supports not only the LTE access but also other 3GPP (GSM/GPRS/EDGE, UMTS/HSPA/HSPA+) and non-3GPP (WLAN (wireless local area network), WiMAX, etc.) access networks.

At the end of 2008 the 3GPP release 8 was completed. It specifies LTE OFDMA-based access and defines the EPC. Because EPC is an all IP network, it does not support voice unless the IP multimedia subsystem (IMS) is implemented.

Two functionalities for the voice service are defined:

- Radio voice call continuity (RVCC). This functionality provides that a VoIP (voice over IP)/IMS service using the LTE radio access moves, if necessary, from the LTE packet switched domain to the 2G or 3G circuit switched domain.
- Circuit switched fallback (CSFB). This functionality does not need IMS and enables circuit switched voice for LTE devices. When an LTE mobile station makes or receives a voice call, it moves ('falls back') to the 3G or 2G network to serve the call.

Home evolved node B (H-eNB) is supported by 3GPP release 8. Release 9 appeared at the end of 2009, introducing enhancements to HSPA and LTE release 8. Release 10 appeared in 2011 and introduced features for LTE advanced.

Figure 1.30 shows the LTE network architecture with its access network, called evolved UTRAN (E-UTRAN), and the evolved packet core (EPC). From the figure is clear that the network structure is extremely simplified. In fact, the only node in the access network is the evolved node B (eNB), which includes the functions of the base station and its controller.

The eNB main functionalities are:

- Radio resource management: radio bearer control, radio admission control, scheduling of resources in uplink and downlink, retransmission handling,
- Mobility management,
- IP header compression and encryption of user data,
- Selection of a mobility management entity (MME) at the UE attachment, if the UE does not provide this information,
- Routing of user data to a gateway,
- Scheduling and transmission of control messages (paging, broadcast), and
- Measurement and measurement reporting configuration for mobility and scheduling.

The eNBs are interconnected through the X2 interface, which carries both data (user plane) and signalling (control plane).

In the EPC, signalling and data are separated and managed by different nodes. The MME is connected to the eNBs through the S1-MME interface, which carries control plane messages. The system architecture evolution gateway (SAE GW) is connected to the eNBs through the S1-U interface, which carries user plane messages. The home subscriber server (HSS) is the repository of all permanent user data, like the subscriber profile and the permanent key for authentication, ciphering and integrity protection. It also stores the location of the user at the level of visited network control node, such as MME.

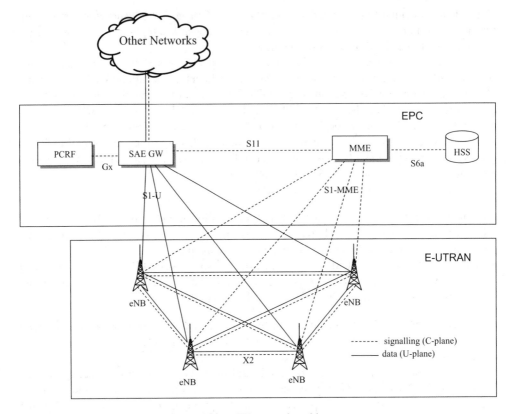

Figure 1.30   LTE network architecture.

The MME manages mobility, MS identities and security parameters. The principal functions of the MME are:

- It executes tracking and paging procedures and assigns temporary identities.
- During the attach (registration) procedure, it receives the attach request message from the eNB and forwards to the SAE GW the 'create default bearer request'.
- It is involved in intra-LTE handover, which includes a core network node reallocation.
- It executes authentication and ciphering procedures.

Contextual with the registration, a connection called the *default bearer* is set up and an IP address is assigned to the MS.

The SAE gateway can be separated into two gateways: the serving gateway (SGW) and the packet gateway (PGW). The primary task of the SGW is IP routing and forwarding of IP packets. It is the user plane anchor for inter-eNB handover and when the user moves among 3GPP access technologies. The PGW is connected to the SGW and to the external networks. It is responsible for the QoS and is the user plane anchor when the MS moves among 3GPP and non-3GPP radio access technologies. One MS can have simultaneous access to more than one PGW (e.g., to have access to different packet data networks). The SAE gateway also performs packet filtering and IP address allocation to mobile stations.

The last element included in the EPC is the policy and charging resource function (PCRF), connected with the Gx interface (signalling only) to the PGW. It implements policy and charging rules and elaborates policy and charging control requests. The EPC can be easily integrated with both 3GPP and non-3GPP access networks, such as fixed broadband, wireless LANs, WiMAX, etc. It allows full mobility among 3GPP access networks and roaming among other access networks. The EPC is an IP-based core network; it does not natively support voice and multimedia.

The initial development of LTE does not support voice service, unless the network operator implements the IP multimedia subsystem (IMS). Release 8 of the 3GPP standard introduces the circuit switched fallback (CSFB) procedure, which provides a mobile station camped on LTE to be served, for the voice service, from the other existing 3G or 2G access networks. This requires 2G/3G coverage and a new interface, the SGW, between the MME and the MSC server.

Figure 1.31 shows the protocol architecture implemented in LTE network nodes. The radio protocol stack in eNBs includes:

- Physical layer (PHY): involves both user and control planes and performs synchronization, channel coding, interleaving, mo-demodulation, multiplexing, measurement and measurement reporting.
- Medium access control (MAC): involves both user and control planes and performs channel access control mechanisms, packet queuing, priority handling. It is the lower part of the second layer in the protocol stack.
- Radio link control (RLC): handles flow control, segmentation, error control, retransmissions. It is the upper part of the second layer in the protocol stack.
- Packet data convergence protocol (PDCP): performs IP header compression for radio transmission in the user plane (UP) and encryption and integrity protection in the control plane (CP).
- Radio resource control (RRC): in the control plane, performs allocation and release of the radio resources, admission and congestion control, and intercell radio resource management. It is at layer 3 in the CP protocol stack.

Other protocols related to the control plane, but not related to the radio interface, are implemented in eNBs and MMEs. Such protocols provide nonaccess stratum (NAS) functionalities and are mobility management (MM), session management (SM), call control (CC) and identity management (IM).

1.2.6 LTE Access Network

The LTE access network is the E-UTRAN, which as a flat architecture of interconnected evolved node Bs (eNBs). The LTE radio interface is based on the following enabling technologies:

- OFDMA (orthogonal frequency division multiple access): used for downlink (from eNB to MS),
- SC-FDMA (single carrier FDMA): used for uplink (from MS to eNB),
- MIMO (multiple input multiple output) antennas,

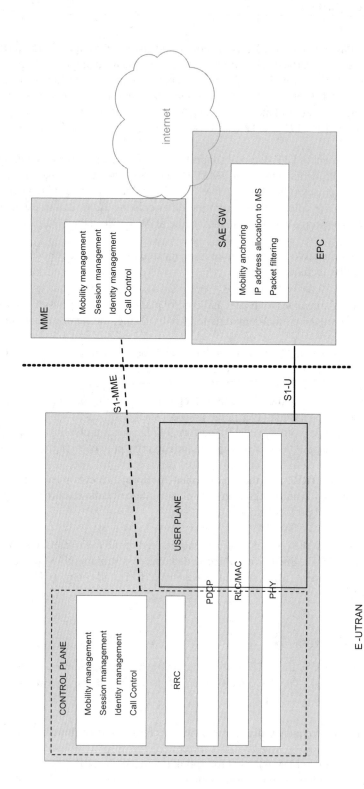

Figure 1.31 LTE protocol architecture.

- Multicarrier channel-dependent resource scheduling, and
- Fractional frequency reuse.

In the following paragraphs are presented the main features of the LTE radio interface.

1.2.6.1 OFDM and OFDMA

OFDMA, which stands for orthogonal frequency division multiple access, is a multiple access technique based on orthogonal frequency division multiplexing (OFDM) modulation, a particular case of multicarrier transmission. In OFDM a high bit rate bit stream is split into low bit rate multiple streams, each of which is transmitted on a separated subcarrier.

OFDM subcarriers (unlike FDM) are formed to partially overlap, allowing a considerable saving in terms of bandwidth. OFDM presents the following advantages:

- The usage of orthogonal subcarriers eliminates noises due to partial spectra overlapping.
- An OFDM signal is the sum of PSK or QAM modulated signals in each subcarrier.
- OFDM is robust in an environment with frequency selective fading.
- OFDM is robust against narrowband interference.
- In slowly time-varying channels, it is possible to adjust the rate of each subcarrier in the function of the signal to noise ratio (SNR) at the receiver side.
- OFDM makes possible single frequency networks.
- OFDM makes possible scalable bandwidth.

In OFDMA, multiple access among different users is achieved by assigning to each user a group of subcarriers in certain slots of time.

Figure 1.32 shows the difference between OFDM and OFDMA: in OFDM each user transmits using all subcarriers; in OFDMA the subcarriers are shared, at the same time, among different users. A user can transmit over contiguous or noncontiguous subcarriers. LTE uses ODFMA for downlink multiple access.

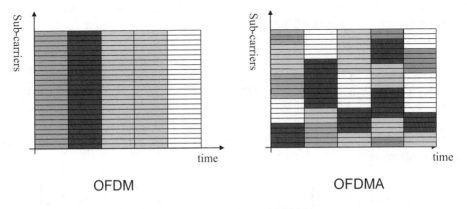

Figure 1.32   OFDM and OFDMA.

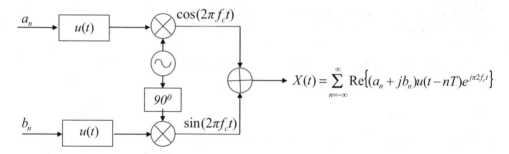

Figure 1.33   QAM modulator scheme.

1.2.6.2   OFDM Basics

OFDM is a multicarrier modulation where a high bit rate bit stream is split into multiple streams at low bit rates, each of which is QAM-modulated on a separated subcarrier. Figure 1.33 shows the QAM-modulator scheme.

The QAM-modulated signal is:

$$X(t) = \sum_{n=-\infty}^{\infty} \mathrm{Re}\left\{(a_n + jb_n)\,u(t - nT)e^{j2\pi f_c t}\right\} \tag{1.5}$$

where T is the symbol time, $(a_n + jb_n)$ is the QAM symbol transmitted at nth symbol time, $u(t)$ is a filter that satisfies the Nyquist condition for the absence of intersymbol interference and f_c is the carrier frequency. A multicarrier signal is generated by summing N QAM modulated signals:

$$X(t) = \mathrm{Re}\left\{\sum_{i=0}^{N-1} X_i(t)\right\} = \mathrm{Re}\left\{\sum_{n=-\infty}^{\infty}\sum_{i=0}^{N-1}\left[X_{n,i}\,u(t - nT)e^{j2\pi f_i t}\right]\right\} \tag{1.6}$$

where T is the symbol time, $X_{n,i} = a_{n,i} + jb_{n,i}$ is the symbol transmitted over the ith subcarrier at nth symbol time and f_i is the ith subcarrier.

Two different subcarriers (i and j) are orthogonal if:

$$\frac{1}{T}\int_{-\infty}^{\infty} u(t)e^{j2\pi f_i t} \cdot u(t)\,e^{j2\pi f_j t}\mathrm{d}t = \begin{cases} 1 & i = j \\ 0 & i \neq j \end{cases} \tag{1.7}$$

In OFDM the subcarriers are spaces of $\Delta f = 1/T$. If $f_i = f_0 + 1/T$, Equation (1.6) can be written as follows:

$$X(t) = \mathrm{Re}\left\{\sum_{i=1}^{N} X_i(t)\right\} = \sum_{n=-\infty}^{\infty} \mathrm{Re}\left\{u(t - nT)e^{j2\pi f_0 t}\sum_{i=0}^{N-1}\left\{X_{n,i}e^{j2\pi t(i/T)}\right\}\right\} \tag{1.8}$$

Observe that the component:

$$X_{bb}(t) = \sum_{i=0}^{N-1} \left\{ X_{n,i} e^{j2\pi t(i/T)} \right\}$$

is the equivalent baseband signal at the nth symbol time and is an inverse Fourier transform.

If $X_{bb}(t)$ is sampled at kT/N intervals, the following discrete sequence is obtained:

$$x_{n,k} = \sum_{i=0}^{N-1} \left\{ X_{n,i} e^{j2\pi i(k/N)} \right\}, \quad k = 0, \ldots, N-1 \tag{1.9}$$

Equation (1.9) is the expression of the inverse discrete Fourier transform (IDFT) of the modulation symbols at time n: $\{X_{n,1} X_{n,2} \ldots X_{n,N}\}$, less than a scale factor N. The sequence $\{x_{n,1} x_{n,2} \ldots x_{n,N}\}$, transmitted in a symbol time T at kT/N intervals, is called the OFDM symbol.

Equation (1.8) suggests an equivalent implementation of a multicarrier modulation with an inverse fast Fourier transform (IFFT) operation in the baseband section of the modulator. The equivalent modulator scheme is shown in Figure 1.34.

At the transmission side, in the symbol time n, N symbols X_i are put as input to the IFFT block to generate the OFDM symbol $\{x_1 \, x_2 \ldots x_N\}$. Then there is a transformation from parallel to serial (at time distance T/N) and, through a digital to analogue converter, two signals (the real and imaginary component of the sequence of $\{x_i\}$) are generated and modulated at frequency f_0.

At the reception side, the f_0 carrier is demodulated and an estimate of the real and imaginary parts of the transmitted signal is derived. The sequence of $\{\hat{x}_i\}$ is obtained by sampling the received signal at step T/N. Then the fast Fourier transform (FFT) operation returns an estimate of the QAM symbol transmitted on each subcarrier. The corresponding demodulator scheme is shown in Figure 1.35.

In an equivalent digital baseband transmission model, a channel can be modelled in the time domain through an FIR (finite impulse response) filter with $v + 1$ taps: $h_0 h_1 \ldots h_v$. If

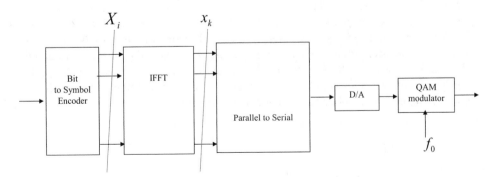

Figure 1.34   OFDM modulator scheme.

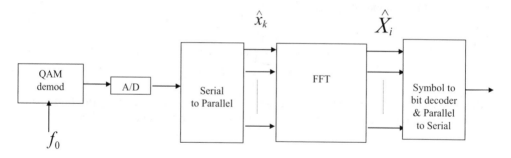

Figure 1.35   OFDM demodulator scheme.

$\{x_1\,x_2\dots x_N\}$ are transmitted on the equivalent baseband channel in a symbol time, the channel output is:

$$x_k = \sum_{i=0}^{v} x_{k-i} h_i$$

The presence of a channel with memory generates inter-OFDM symbols interference. In order to eliminate such interference, a prefix of v symbols is added to the transmitted sequence of $\{x_i\}$. The periodicity of the input sequence is simulated by replicating the last v samples of the sequence and placing them in the head. The obtained prefix is called the cyclic prefix. Figure 1.36 shows the equivalent baseband transmission scheme with the insertion of the cyclic prefix at the transmission side and removal of added samples at the reception side.

1.2.6.3   SC-FDMA Basics

Figure 1.37 shows the scheme of a single carrier FDMA (SC-FDMA) transmission. Unlike OFDM, the QAM symbols are turned to the frequency domain through a discrete Fourier transform (DFT) block, obtaining a sequence of discrete symbols that are associated with the subcarriers and then converted back in time through an IFFT.

    The SC-FDMA is used in LTE for uplink multiple access. It is based on FDMA but it implies the use of orthogonal subcarriers. A DFT preprocessing is added to a conventional OFDMA transmitter. The frequency resource is shared among different users by assigning nonoverlapping adjacent groups of subcarriers. Therefore, bandwidth allocation in uplink is continuous. The reason for choosing SC-FDMA for the uplink is that it presents a low peak-to-average power ratio (PAPR), which implies an efficient use of power amplifiers and therefore a significant reduction in mobile station battery consumption.

1.2.6.4   MIMO Basics

A MIMO system consists of m transmit and n receive antennas, with $n \geq m$. If a narrowband channel is assumed, the connection between the transmitting antenna i and the receiving

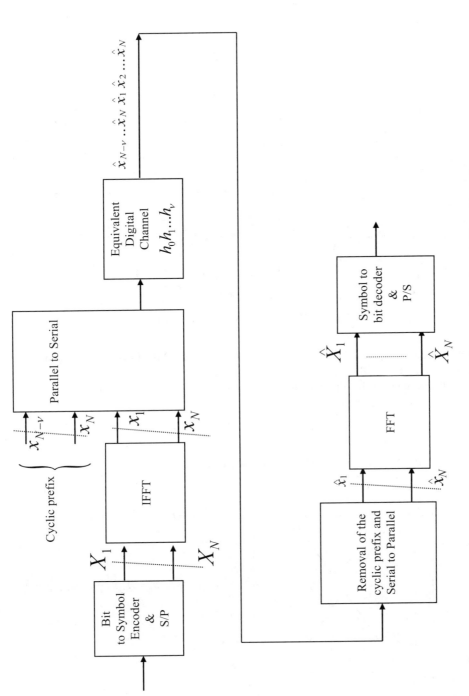

Figure 1.36 Equivalent OFDM baseband transmission scheme with cyclic prefix.

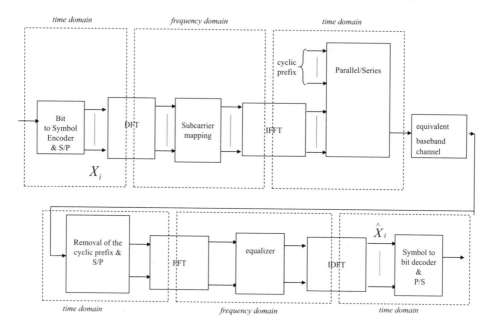

Figure 1.37 SC-FDMA transmission scheme.

antenna j can be expressed through a component h_{ij} and the MIMO system can be modelled as shown in Figure 1.38.

If s_i is the signal at the ith transmitting antenna ($i = 1, 2, \ldots, m$), the signal at the jth ($j = 1, 2, \ldots, n$) receiving antenna is:

$$y_j = h_{j1}s_1 + h_{j2}s_2 + \cdots + h_j s_m + n_j \tag{1.10}$$

where n_j is the received noise component at the jth antenna.

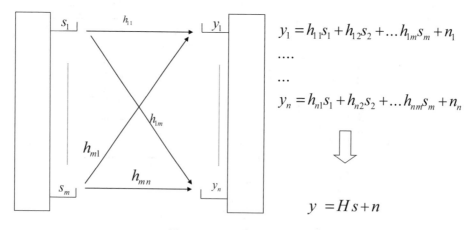

Figure 1.38 MIMO model.

In a MIMO system three techniques live together:

- Diversity
- Spatial multiplexing
- Beamforming

Diversity

The diversity technique can be applied at both the transmitter side and the receiver side. The purpose of diversity is to improve the quality of the received signal, namely having a higher signal to noise ratio (SNR) at the receiver side.

Because of different transmission paths, the receiver sees differently faded signals. The most used techniques for the diversity at the receiver side are:

- Switched diversity, which always chooses the strongest signal,
- Equal gain combining (EGC), which coherently sums the received signals, and
- Maximum ratio combining (MRC), which optimally combines the received signals, where the signals are weighted proportionally to their signal to noise ratio.

The receive diversity does not require the compliance to a standard.

In the diversity at the transmission side, copies of the same signal, differently encoded, are sent through the transmitting antennas. Space–time codes are used for transmit diversity, where multiple copies of the signal are transmitted from different antennas at different times. Alamouti developed a two-branch transmit diversity scheme with two transmitting and two receiving antennas [7]. Pseudo-Alamouti coding schemes have been developed for multiple antennas.

Beamforming

Beamforming is a technique that creates a directional radiation pattern to a user, thus reducing the interference and increasing the antenna gain. The final goal is to increase the SNR at the receiver side.

There are two different beamforming techniques:

- Switched beamforming is based on phased array antennas with many fixed predefined radiation patterns. The radiation pattern is chosen as a function of the requirements of the cell (i.e., position of users) and can be varied if conditions change.
- Adaptive beamforming is based on adaptive array antennas, able to adjust the radiation pattern as a function of the user's position. With adaptive beamforming it is possible to have a null of the radiation pattern corresponding to the interfering signals, thus optimizing the SNR at the receiver side.

Figure 1.39 shows two examples of switched and adaptive beamforming.

Spatial Multiplexing

The goal of spatial multiplexing is to increase the data rate rather than improve the quality of the transmission. The bit rate is improved by a factor m through the transmission of different streams via separate antennas. In general, spatial multiplexing requires low correlation among

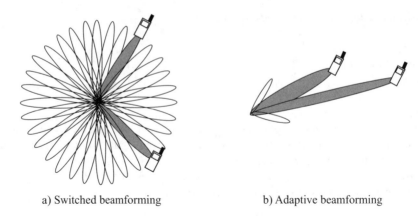

a) Switched beamforming   b) Adaptive beamforming

Figure 1.39   Switched and adaptive beamforming.

propagation paths, which implies that matrix **H** is full rank (i.e., the row vectors are linearly independent).

The streams can be decoupled at the receiver side by using the following different methods:

- Open loop: other than the data stream, a known sequence is transmitted for channel estimation. The receiver can apply interference cancellation techniques.
- Closed loop: to decouple the propagation paths a feedback information, called channel state information (CSI), is sent from the receiver to the transmitter. A precoding matrix **W** is applied at the transmission side according to the estimated channel matrix. Figure 1.40 shows an example where the channel matrix **H** is decomposed on eigenvalues ($\mathbf{H} = \mathbf{U\Sigma V}^{-1}$, where $\mathbf{\Sigma}$ is the diagonal eigenvalues matrix, $\mathbf{U}^{-1}\mathbf{U} = \mathbf{I}$ and $\mathbf{V}^{-1}\mathbf{V} = \mathbf{I}$). In this case the precoding matrix is $\mathbf{W} = \mathbf{V}$. At the receiver, by applying $\mathbf{U}^{-1}$ the propagation paths are perfectly decoupled. In fact, the signal received on the ith antenna is $y_j = \sigma_j s_j$.

The choice of the correct MIMO technique strongly depends on the channel estimation. If the channel matrix **H** is low-rank (i.e., mobile channel), then diversity techniques are used; if the channel matrix **H** is high-rank (i.e., static channel), then spatial multiplexing with a number of spatial streams equal to the matrix rank is preferred.

Collaborative MIMO (Multiuser MIMO)
In multiuser MIMO (MU-MIMO), the idea is that different special streams belong to different users. This technique is useful in uplink because it needs only one transmitting antenna.

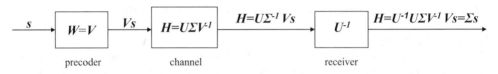

precoder   channel   receiver

Figure 1.40   Example of closed loop spatial multiplexing.

Table 1.6   Bands for LTE FDD

Band	Uplink (MHz)	Downlink (MHz)
2100	1920–1980	2110–2170
1900	1850–1910	1930–1990
1800	1710–1785	1805–1880
1700/2100	1710–1755	2110–2155
850	824–849	869–894
800	830–840	875–885
2600	2500–2570	2620–2690
900	880–915	925–960
1700	1750–1785	1845–1880
1700/2100	1710–1770	2110–2170
1500	1427.9–1452.9	1475.9–1500.9
US700	698–716	728–746
US700	777–787	746–756
US700	788–798	758–768
US700	704–716	734–746
Japan800	815–830	860–875
Japan800	830–845	875–890

1.2.6.5   LTE Radio Interface

LTE is specified in both TDD and FDD modes, with a scalable bandwidth. The specified bandwidths are in MHz: 1.4, 3, 5, 10, 15 and 20. Some of the possible bands for LTE FDD are listed in Table 1.6. Some of the possible bands for LTE TDD are listed in Table 1.7.

In the time domain, a frame of 10 ms is defined in both TDD and FDD modes, with slots of 0.5 ms. Also one subframe of 1 ms, which is the transmission time interval (TTI), is defined. In TDD it is possible to dynamically change the uplink and downlink allocation in order to meet load requirements. The multiple access technique is OFDMA in downlink and SC-FDMA in uplink, both based on subcarrier spacing of 15 kHz, regardless of the channel bandwidth. The number of subcarriers varies from 72 in a channel of 1.4 MHz to 1200 in a 20 MHz channel.

The atomic radio resource is the *resource element*, which is one OFDM symbol in one subcarrier. The radio resource is the *physical resource block*, which includes 12 subcarriers in

Table 1.7   Bands for LTE TDD

Band	Uplink and downlink (MHz)
UMTS TDD1	1900–1920
UMTS TDD2	2010–2025
US1900 UL	1850–1910
US1900 DL	1930–1990
US1900	1910–1930
2600	2570–2620
UMTS TDD	1880–1920
2300	2300–2400

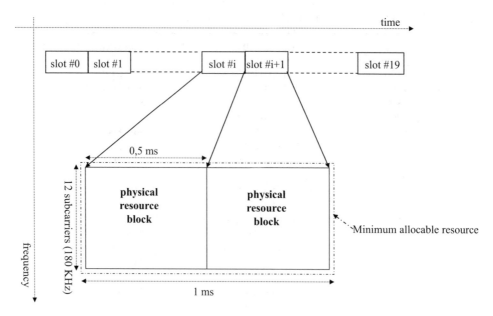

Figure 1.41 LTE frame structure and resource blocks.

a slot of 0.5 ms. Therefore, a physical resource block occupies a bandwidth of 180 MHz in 0.5 ms. This is valid for both downlink (OFDMA) and uplink (SC-FDMA). However, because TTI in LTE is 1 ms, the minimum allocable resource is a group of 12 subcarriers (180 MHz) in a time of 1 ms. Figure 1.41 shows the frame with 20 time slots, the resource block and the minimum allocable resource in the TTI of 1 ms.

Table 1.8 shows the different number of subcarriers and resource blocks in the case of channel bandwidths of 1.4, 3, 5, 10 and 20 MHz. Tables 1.9 and 1.10 show some downlink and uplink modulation and coding schemes with corresponding bit rates at the physical layer as a function of the bandwidth (expressed in Mbps). The modulation and coding scheme can be changed per allocated resource and is adaptive with respect to the radio channel conditions. MIMO usage can be single stream (SS) or multiple streams (2×2 or 4×4). In the case of multiple streams, spatial multiplexing is used to multiply the bit rate.

1.2.6.6 LTE Physical, Transport and Logical Channels

Figure 1.42 shows the LTE radio protocol stacks for user and control planes. The radio protocol stacks for user and control planes include the physical layer, medium access control (MAC),

Table 1.8 Number of subcarriers and resource blocks for different channel bandwidths

Channel bandwidth (MHz)	1.4	3	5	10	15	20
Number of subcarriers	72	180	300	600	900	1200
Number of resource blocks	6	15	25	50	75	100

Table 1.9 Downlink modulation and coding schemes with corresponding bit rates (Mbps)

Modulation	Coding rate *k/n*	MIMO usage	Bandwidth					
			1.4 MHz	3.0 MHz	5.0 MHz	10 MHz	15 MHz	20 MHz
QPSK	1/2	SS	0.8	2.2	3.7	7.4	11.2	14.9
16-QAM	1/2	SS	1.5	4.4	7.4	14.9	22.4	29.9
16-QAM	3/4	SS	2.3	6.6	11.1	22.3	33.6	44.8
64-QAM	3/4	SS	3.5	9.9	16.6	33.5	50.4	67.2
64-QAM	1	SS	4.6	13.2	22.2	44.7	67.2	89.7
64-QAM	3/4	2 × 2	6.6	18.9	31.9	64.3	96.7	129.1
64-QAM	1	2 × 2	8.8	25.3	42.5	85.7	128.9	172.1
64-QAM	1	4 × 4	16.6	47.7	80.3	161.9	243.5	325.1

Table 1.10 Uplink modulation and coding schemes with correspondent bit rates (Mbps)

Modulation	Coding rate *k/n*	MIMO usage	Bandwidth					
			1.4 MHz	3.0 MHz	5.0 MHz	10 MHz	15 MHz	20 MHz
QPSK	1/2	SS	0.9	2.2	3.6	7.2	10.8	14.4
16-QAM	1/2	SS	1.7	4.3	7.2	14.4	21.6	28.8
16-QAM	3/4	SS	2.6	6.5	10.8	21.6	32.4	43.2
16-QAM	1	SS	3.5	8.6	14.4	28.8	43.2	57.6
64-QAM	3/4	SS	3.9	9.7	16.2	32.4	48.6	64.8
64-QAM	1	SS	5.2	13.0	21.6	43.2	64.8	86.4

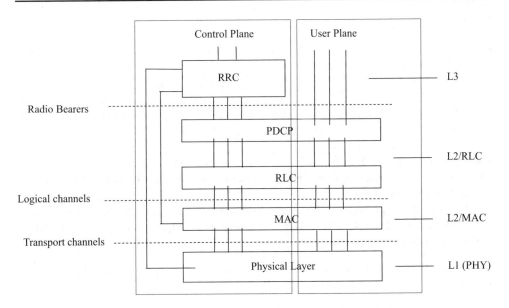

Figure 1.42 LTE radio protocol stacks for user and control planes.

radio link control (RLC) and packet data convergence protocol (PDCP). Radio resource control (RRC) is on top of PDCP for the control plane. The physical layer is responsible for all the radio transmission functions.

MAC, RLC and PDCP are layer two protocols. MAC multiplexes packet data units (PDUs) from one or more logical channels into a transport channel. The RLC offers link control over the radio interface for user and control data. PDCP provides services in the form of radio bearers to RRC for the control plane and to the IP layer for the user plane. Radio bearers are services used to deliver C-plane and U-plane over the radio interface. RRC is responsible for layer three signalling exchange between the mobile station and the evolved node B.

Logical channels, which are the resources offered from the MAC layer to the upper layers, are characterized by the nature of their information content; they carry information related to both the user and control plane. Logical channels are:

- Broadcast control channel (BCCH): in downlink, broadcasts system information
- Paging control channel (PCCH): in downlink, carries paging
- Common control channel (CCCH): uplink and downlink, used to transmit Radio Resource control (RRC) initial bidirectional signalling between the MS and eNB
- Dedicated control channel (DCCH): uplink and downlink, used to transmit dedicated RRC signalling between the MS and eNB
- Dedicated traffic channel (DTCH): uplink and downlink, dedicated to user data transmission
- Multicast control channel (MCCH): point-to-multipoint, downlink, for multicast control information
- Multicast traffic channel (MTCH): point-to-multipoint, downlink, for multicast traffic

Transport channels are resources offered from the physical to the MAC layer. Transport channels in LTE are shared channels: there are no dedicated transport channels. Transport channels are:

- Broadcast channel (BCH): in downlink, carries part of the system information (SI); additional SI blocks are mapped in the DL-SCH.
- Paging channel (PCH): in downlink, carries paging information.
- Downlink shared channel (DL-SCH): is the most important resource in downlink. Carries data (DTCH) and signalling (BCCH, CCCH and DCCH). Can also carry multicast information. Supports hybrid automatic repeat request (HARQ), dynamic packet scheduling, adaptive modulation and coding.
- Multicast channel (MCH): downlink, carries multicast information.
- Random access channel (RACH): uplink, carries control information between the MS and eNB. Is subject to collisions.
- Uplink shared channel (UL-SCH): is the most important resource in uplink. Carries data (DTCH) and signalling (CCCH and DCCH). Supports HARQ, dynamic packet scheduling, adaptive modulation and coding.

Physical channels are the resources used for transmission over the radio interface. Physical channels are:

- Physical broadcast channel (PBCH): carries the master information block (MIB) from the BCH. In LTE, MIB contains very limited information, like the cell bandwidth, the physical hybrid ARQ indicator channel (PHICH) structure, the system frame number (SFN).

- Physical HARQ indicator channel (PHICH): downlink, sends ACK/NACKs related to the uplink transmissions on the physical uplink shared channel (PUSCH). Each PHICH is addressed to one MS.
- Physical downlink shared channel (PDSCH): is the most important downlink resource and carries data and signalling. Is allocated on a TTI basis (1 ms) to the mobile stations. The MAC scheduler assigns channel coding, modulation, subcarrier allocation and addresses DL transmissions in the PDCCH.
- Physical downlink control channel (PDCCH): downlink, is used to assign resources in the PDSCH (downlink) and in the PUSCH/PUCCH (uplink).
- Physical control format indicator channel (PCFICH): downlink, indicates dimension, in OFDM symbols, of the control region used for PDCCH in the same subframe.
- Physical multicast channel (PMCH): downlink, is used for multicast.
- Physical uplink shared channel (PUSCH): is the most important uplink resource in a cell and carries data and signalling. Is allocated on a TTI basis (1 ms) to the mobile stations. The MAC scheduler assigns, per each MS, channel coding and modulation, and allocates the subcarriers.
- Physical uplink control channel (PUCCH): carries a channel quality indicator (CQI) and ACK/NACK.
- Physical random access channel (PRACH): carries random access preambles during the random access procedure.

Figure 1.43 shows downlink and uplink channels at logical, transport and physical layers and their mapping. In the figure are also shown the following signals:

- Primary synchronization signal (PSS) and secondary synchronization signal (SSS): both in downlink, are used for cell search and identification by the MS. Together they carry the cell identifier (Id).
- Reference signals (RSs): used in uplink and downlink for channel estimation.

When an MS switches on, it searches and selects one cell through the synchronization signals, then derives system information from the PBCH, executes the random access procedure and sends signalling to register to the network and to set up the default bearer.

The assignment of downlink radio resources in the function of the radio conditions is performed through the *channel feedback reporting procedure*. During the eNB transmission, the MS measures the downlink channel and sends a feedback to the eNB, which can include the following parameters:

- Channel quality indicator (CQI): the MS indicates the transmission format (modulation and coding scheme and transport block size, or TBS) that the mobile can receive in the next TTI with a block error rate (BLER) lower than 10%.
- Rank indicator (RI): for MIMO, indicates the rank of the estimated channel matrix **H**. Equals the number of usable spatial streams. It is related to the whole system bandwidth.
- Precoding matrix indicator (PMI): indicates the preferred precoding matrix in a codebook.

Table 1.11 shows the considered parameters for uplink feedback. Based on the feedback, the eNB assigns downlink resources to the MS.

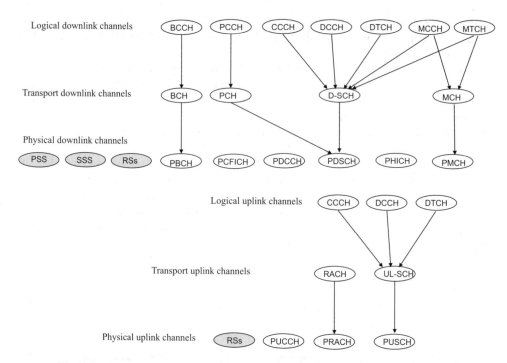

Figure 1.43   Downlink and uplink channels at logical, transport and physical layers.

Figure 1.44 shows the messages exchanged between an MS and an eNB with the corresponding physical channels during the phases of synchronization and system nformation acquisition, random access, contention resolution, NAS message transfer, downlink data transfer and uplink data transfer.

1.2.7   LTE Advanced

LTE advanced is the evolution of LTE, standardized starting from release 10 of the 3GPP. The effort was to align release 10 with the requirements set up by the ITU for an IMT advanced system. Table 1.12 shows a comparison between the IMT requirements for a 4G system and the corresponding performances of LTE releases 8 and 10 [8].

Table 1.11   Uplink feedback

		Frequency selective	Available in open loop	Available in closed loop
PMI	Index of MIMO precoding matrix in the codebook	Yes	No	Yes
RI	Number of MIMO supported data streams	No	Yes	Yes
CQI	Supported transmission format	Yes	Yes	Yes

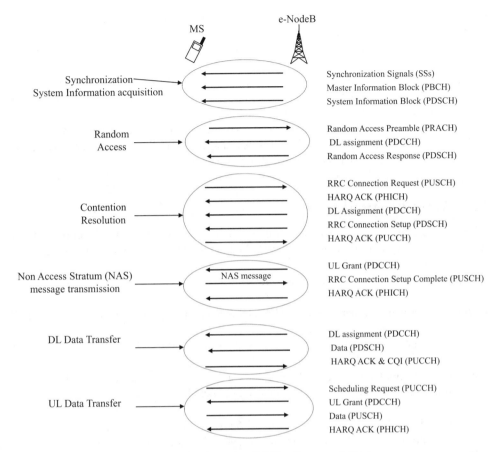

Figure 1.44   Example of LTE radio transmission.

The principal radio features introduced in LTE advanced are:

- Carrier aggregation: two or more carriers, each with a bandwidth of up to 20 MHz, are aggregated. Contiguous and noncontiguous carriers can be aggregated, up to 100 MHz.
- Extended MIMO configurations: up to an 8×8 MIMO configuration is considered in downlink; up to a 4×4 MIMO configuration is considered in uplink.

Table 1.12   IMT requirements and LTE release 8 and 10 performances

	IMT requirements	LTE release 8	LTE release 10
Peak DL data rate	1 Gbps	$\sim$ 325 Mbps	> 1 Gbps
Peak UL data rate	500 Mbps	$\sim$ 85 Mbps	> 500 Mbps
Control plane latency	50 ms	50 ms	50 ms

- Coordinated multiple point (CoMP) transmission and reception: this feature improves high data rate coverage and cell-edge throughput. Downlink coordinated multipoint transmission introduces dynamic coordination among multiple geographically separated transmission points.
- Relaying: used to improve the high data rates coverage and the cell-edge throughput, and to provide coverage in new areas. The relay node is wirelessly connected to a radio access network via a donor cell. There are two bidirectional radio links/interfaces: the *access link*, between the MS and the relay node (RN), and the *backhaul link*, between the RN and the eNB. The relaying can be *inband*, if access and backhaul link use the same band, and *outband*, if access and backhaul link use different bands.

1.3   Wireless Networks

Wireless networks are access networks with a radio connection between the mobile terminal and the first access network node. In general, they do not handle user mobility and therefore are not radiomobile networks.

Wireless networks can be grouped as follows:

- Wireless personal area network (WPAN): are very short range networks, for low or high data rate transmission.
- Wireless local area network (WLAN): are local area networks with a radio interface. Their range is of tens, or hundreds, of metres.
- Wireless metropolitan area network (WMAN): are networks with coverage of the order of few kilometres. They are used mainly for wireless local loop connectivity.

Figure 1.45 shows different coverage ranges of WPANs, WLANs and WMANs.

Wireless networks are different for coverage range, supported bit rates, standards, etc. Figure 1.46 shows the principal IEEE standards for wireless networks.

1.3.1   Wireless LAN

A wireless local area network (WLAN) is a local area network with a wireless connection between the user and the network. WLANs can be used for indoor areas (i.e., offices, homes,

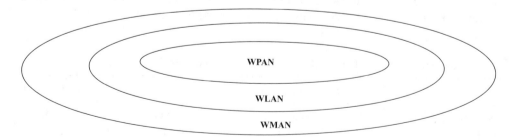

Figure 1.45   Coverage ranges of WPANs, WLANs and WMANs.

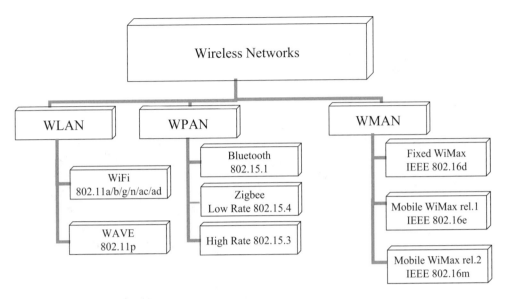

Figure 1.46 Principal IEEE standards for wireless networks.

hotels) or limited outdoor areas (i.e., campuses). The principal standard for WLANs is IEEE 802.11, but there is also HiperLAN, the European standard for WLANs.

HiperLAN proposes a solution for local wireless IP transport, defining a physical layer based on the frequency shift keying (FSK) modulation, and operating in the unlicensed band around 5 GHz. HiperLAN/2 extends the first type and is proposed for point-to-point and point-to-multipoint connections. It uses OFDM modulation, TDD duplexing and TDMA multiple access. It also provides quality of service (QoS). Limits on the maximum powers are fixed from the regulator bodies.

IEEE 802.11 defines two network topologies:

- Infrastructure: the WLAN is used to extend a fixed network infrastructure (i.e., LAN, ADSL, or asymmetric digital subscriber line).
- Ad hoc: the stations communicate through a direct link to form an ad hoc network.

Figure 1.47 shows an example of infrastructure and ad hoc network topologies.

The infrastructure network consists of an access point (AP) connected to the network and a set of wireless stations (WSs). The AP with the associated WSs forms a basic service set (BSS). A set of two or more BSSs forming a single network is an extended service set (ESS). An ad hoc wireless network forms an independent basic service set (IBSS).

1.3.1.1   IEEE 802.11a, b and g

IEEE 802.11 defines a common multiple access control for different physical layers b, a and g. Table 1.13 shows the principal characteristics of b, a and g transmission standards [9]. The

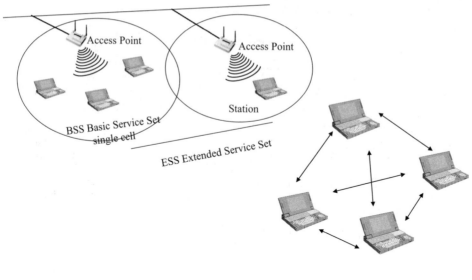

Independent Basic Service Set
(IBSS)

Figure 1.47   Examples of infrastructure and ad hoc mode.

first row shows that b and a standards were approved in 1999, while the g standard came out in 2003, and includes b. IEEE 802.11b and g work on the unlicensed 2.4 GHz frequency, while IEEE 802.11a uses the unlicensed 5 GHz band. The maximum bit rate at the physical layer is 11 Mbps for IEEE 802.11b and 54 Mbps for IEEE 802.11a and g. The radio channel is about 20 MHz wide. All the WLAN standards implement a link adaptation mechanism that adjusts the transmission format (modulation and coding) to the estimated radio link quality.

Table 1.13   Principal characteristics of b, a, and g transmission standards

	802.11b	802.11a	802.11g
Standard approval	July 1999	July 1999	June 2003
Maximum physical layer bit rate	11 Mbps	54 Mbps	54 Mbps
Modulation	CCK	OFDM	OFDM and CCK
PHY bit rates	1, 2, 5.5, 11 Mbps	6, 9, 12, 18, 24, 36, 48, 54 Mbps	CCK: 1, 2, 5.5, 11 Mbps OFDM: 6, 9, 12, 18, 24, 36, 48, 54 Mbps
Frequencies	2.4–2.497 GHz	5.15–5.35 GHz 5.425–5.675 GHz 5.725–5.875 GHz	2.4–2.497 GHz

Reproduced with permission from Kaveh Pahlavan, Allen H. Levesque, "Wireless Information Networks," John Wiley & Sons, 2005.

Table 1.14 Low bit rates (1 and 2 Mbps) encoding for IEEE 802.11b

Bit rate	Code length	Modulation
1 Mbps	11	BPSK
2 Mbps	11	QPSK

IEEE 802.11b uses the direct sequence spread spectrum (DSSS), with speeds of 1, 2, 5.5 and 11 Mbps. Table 1.14 shows that 1 and 2 Mbps are obtained by changing the modulation (BPSK or QPSK) and by maintaining the binary Barker code for the spread spectrum. The Barker code is the following spreading sequence of 11 chips: 1–111–1111–1–1–1.

Complementary code keying (CCK) is the encoding technique adopted for 5.5 and 11 Mbps, and the principle is shown in Figure 1.48. The CCK codeword is:

$$CCK_{codeword} = \left\{ e^{j(\phi_1+\phi_2+\phi_3+\phi_4)}, e^{j(\phi_1+\phi_3+\phi_4)}, e^{j(\phi_1+\phi_2\cdot\phi_4)}, -e^{j(\phi_1+\phi_4)}, e^{j(\phi_1+\phi_2+\phi_3)}, e^{j(\phi_1+\phi_3)}, \right.$$
$$\left. -e^{j(\phi_1+\phi_2)}, e^{j(\phi_1)} \right\}$$

$$(1.11)$$

Different bit-phase mapping is used for 5.5 and 11 Mbps bit rates. In the first case, four bits are mapped into four phases to obtain the CCK codeword; in the second case, eight bits are mapped into four phases to obtain the CCK codeword. The codeword identifies the QPSK modulation symbols.

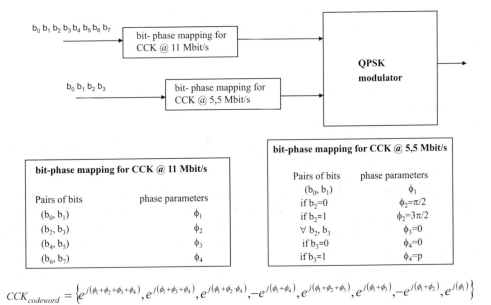

Figure 1.48 High bit rates (5.5 and 11 Mbps) encoding for IEEE 802.11b.

Table 1.15   IEEE 802.11a and IEEE 802.11g modulation and coding schemes

PHY bit rate (Mbps)	Coding rate	Modulation
6	1/2	BPSK
9	3/4	BPSK
12	1/2	QPSK
18	3/4	QPSK
24	1/2	16-QAM
36	3/4	16-QAM
48	2/3	64-QAM
54	3/4	64-QAM

The radio channels in IEEE 802.11 are about 20 MHz wide, and in general are not separated but overlapping. This results in a poorer radio link and then in a lower bit rate per user. IEEE 802.11g and IEEE 802.11a introduce orthogonal frequency division multiplexing (OFDM), with transmission formats (modulation and coding) adapted to the radio link quality. The modulation and coding schemes are the same for the two standards g and a, and are shown in Table 1.15.

1.3.1.2   IEEE 802.11 Medium Access Control

In the infrastructure mode, the wireless stations (WSs) must be synchronized to a common clock, distributed from the access point (AP). The AP transmits, at target beacon transmission (TBTT) intervals, the beacon frame, where it copies the value of its timer (*timestamp*).The stations of the BSS update their timer at that value. The other values broadcasted in the beacon frame are:

- Beacon interval;
- Service set identifier (SSId); the AP can be configured not to transmit the SSId;
- Supported rates;
- Physical layer parameter set; and
- Traffic information map: carries the identifiers of the WSs having data to be transmitted in the next TBTT and is used for power save.

The MAC layer in IEEE 802.11 includes three functions: the distributed coordination function (DCF), DCF-request to send/clear to send (DCF-RTS/CTS) and point coordination function (PCF). The DCF is mandatory in IEEE 802.11. It uses carrier sense multiple access–collision avoidance (CSMA/CA). Before transmission, each WS listens to the medium; if the channel is sensed free for an established interval, called the distributed interframe space (DIFS), then the station transmits; otherwise it executes a backoff procedure, which calculates a backoff time through a random function. The backoff time is calculated as follows:

$$\text{Backoff\_time} = \text{Random()} \times \text{slot\_time} \qquad (1.12)$$

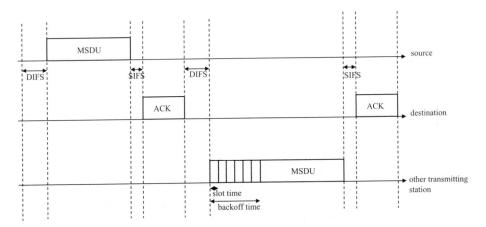

Figure 1.49   Example of CSMA/CA mechanism with two wireless stations.

Equation (1.11) assumes, according to a uniform distribution, values in the range [0, CW], where CW is the contention window. The slot time is an interval strictly dependent on the physical layer (i.e., 20 μs for 802.11b and 9 μs for 802.11a). As soon as the wireless station senses the medium is free, it starts to decrease the backoff timer and freezes the countdown if the radio becomes busy. When the counter reaches the value zero, the WS, after having sensed the medium free for a DIFS, transmits the MAC service data unit (MSDU). If the MSDU is correctly received, it is acknowledged through an ACK message. If the ACK is not received in a short interframe space (SIFS), the MSDU is retransmitted. The SIFS is shorter than the DIFS. Figure 1.49 shows an example of the CSMA/CA mechanism with two wireless stations.

An optional function provided from the IEEE 802.11 MAC is a distributed coordination function-request to send/clear to send (DCF-RTS/CTS), also called a virtual carrier sense (VCS). The virtual carrier sense solves the 'hidden node' problem, shown in Figure 1.50.

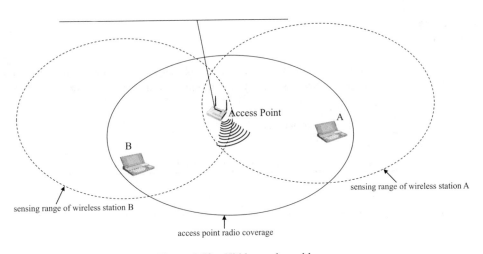

Figure 1.50   Hidden node problem.

In the figure, WSs A and B see the access point, but A is not able to see B and vice versa. A is a hidden node for B and B is a hidden node for A. The mechanism of a two-way handshake, realized through the introduction of the packet's request to send (RTS) and clear to send (CTS), solves the problem of hidden nodes. With virtual carrier sense wireless station A, following the rules of DCF, transmits an RTS packet. The RTS contains the source address, the destination address and the duration of the transmission expressed in µs. The access point acknowledges the reception of the RTS through a clear to send (CTS) packet, where it copies the duration field of the received RTS, minus the value of an SIFS and of the time needed to transmit the CTS frame:

$$\text{Duration\_Rx} = \text{Duration\_Tx} - t(\text{CTS}) - \text{SIFS}$$

The CTS is received from all the stations in the AP range. After receiving the RTS or CTS packet, a station sets the network allocation vector (NAV) timer at the value 'duration' contained in the received packet, and it will not attempt the access the medium until the NAV reaches the value zero. With virtual carrier sense collisions occur only among RTS/CTS packets, which are very short.

The point coordination function (PCF) is optional in IEEE 802.11 and is provided only in the infrastructure mode. In this case the access is controlled from the AP, acting as the point coordinator (PC). Contention free transmissions are allowed in a period called the contention free period (CFP). CFP begins with a beacon frame and alternates to the contention period (CP), where the access is regulated from the DCF. The alternation of CFP and CP is shown in Figure 1.51. A new time interval, the PCF interframe space (PIFS), which is smaller than the DIFS, is introduced to separate the end of the CP and the beacon transmission.

1.3.1.3   IEEE 802.11n

In 2009, IEEE 802.11n final standard version was approved. It introduces enhancements to the MAC layer to decrease the latency and to the physical layer to improve the bit rate. The main physical layer enhancements are:

- Dual carrier
- MIMO

The maximum physical bit rate achievable in IEEE 802.11n is 600 Mbps using a 40 MHz wide channel and four spatial streams. Thirty-one modulation and coding schemes (MCSs)

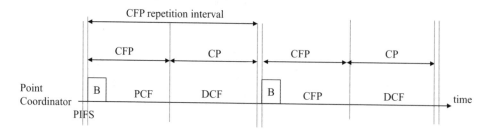

Figure 1.51   Alternation of DCF and PCF.

Table 1.16 IEEE 802.11n modulation and coding schemes, with carrier aggregation

Modulation	Coding rate	Number of spatial streams	PHY bit rate (Mbps)			
			20 MHz channel		40 MHz channel	
			GI 800 ns	GI 400 ns	GI 800 ns	GI 400 ns
BPSK	1/2	1	6.5	7.2	13.5	15
QPSK	1/2	1	13	14.4	27	30
QPSK	3/4	1	19.5	21.7	40.5	45
16-QAM	1/2	1	26	28.9	54	60
16-QAM	3/4	1	39	43.3	81	90
64-QAM	2/3	1	52	57.8	108	120
64-QAM	3/4	1	58.5	65	121.5	135
64-QAM	5/6	1	65	72.2	135	150
BPSK	1/2	2	13	14.4	27	30
QPSK	1/2	2	26	28.9	54	60
QPSK	3/4	2	39	43.3	81	90
16-QAM	1/2	2	52	57.8	108	120
16-QAM	3/4	2	78	86.7	162	180
64-QAM	2/3	2	104	115.6	216	240
64-QAM	3/4	2	117	130	243	270
64-QAM	5/6	2	130	144.4	270	300
BPSK	1/2	3	19.5	21.7	40.5	45
....						
64-QAM	5/6	4	260	288.9	540	600

are defined in IEEE 802.11n, and some of them are shown in Table 1.16. The bit rates change if the guard interval (GI), which is the duration of the cyclic prefix, is 400 or 800 ns.

Under development are two other standards, IEEE 802.11ac and IEEE 802.11ad. IEEE 802.11ac and IEEE 802.11ad provide high throughput respectively on the 5 GHz and 60 GHz bandwidths, up to Gbps.

1.3.1.4 IEEE 802.11p

IEEE 802.11p provides some changes to the IEEE 802.11 standard to fasten link setup and ad hoc mode operation. The physical layer remains almost unchanged, except for the frequencies, which are in the spectrum of 5.9 GHz. IEEE 802.11p is considered for wireless access in vehicular environments (WAVEs) and can be the wireless interface for intelligent transportation systems (ITSs).

In terms of functional requirements, the ITS architecture provides a complete interaction among:

• Service centers, which manage the ITS applications, are IP interconnected to the other entities.

• Next generation roadside infrastructure, capable of interworking with the already deployed infrastructure, is connected to the service centres and realizes the short–medium range wireless communication with the vehicles.
• Vehicular infrastructure, capable of supporting communications with the other vehicles and with roadside infrastructure.
• Personal devices used by the driver or the passenger to communicate with the ITS world. These elements are interconnected and exchange data and information through a set of wireless communication networks.

1.3.2   Wireless MAN

A wireless metropolitan area network (WMAN) is a wireless network with a wide coverage, like a metropolitan area. The principal standard for WMANs is IEEE 802.16, which has two versions:

• IEEE 802.16d, standardized in 2004 and called fixed WiMax.
• IEEE 802.16e, which came out in 2005 as an amendment to the IEEE 802.16d. Even if IEEE 802.16e does not automatically imply mobility, it is often identified with mobile WiMax.

The term WiMAX is often used to identify IEEE 802.16 standard systems, but it comes from the WiMAX forum, an organization that certifies and promotes the compatibility and interoperability of broadband wireless products based on IEEE Standard 802.16.

In many countries fixed WiMAX works around 3.5 GHz. General characteristics of fixed WiMAX are:

• Duplexing can be TTD or FDD.
• The physical layer adopts OFDM modulation, with QPSK, 16-QAM, 64-QAM in uplink and downlink.
• The choice of the modulation and coding scheme is adapted to the radio channel conditions.
• The bandwidth is variable, but with a fixed number of 256 subcarriers.
• The MAC layer is connection oriented, based on TDMA in uplink and downlink. It supports quality of service (QoS).

1.3.2.1   IEEE 802.16e

Mobile WiMAX [10] implies the idea of mobility. In that case, it must include procedures for mobility handling and handover, and can be counted among the third generation systems.

The network architecture of mobile WiMAX is shown in Figure 1.52. It is an all IP network architecture, with support of QoS. The mobile station (MS) connects, through a standard radio interface, with a mobile WiMax base station (BS). The BS realizes the radio coverage, broadcasts cell parameters, makes measurements over the uplink, receives the channel quality indicators (CQIs) from the MSs and schedules radio resources to the mobile users. The base stations are connected through the R8 interface. The mobile WiMAX base stations are also connected, through the R6 interface, to the access service network gateway (ASN-GW), which manages the user and control planes and connects the WiMAX access network to the

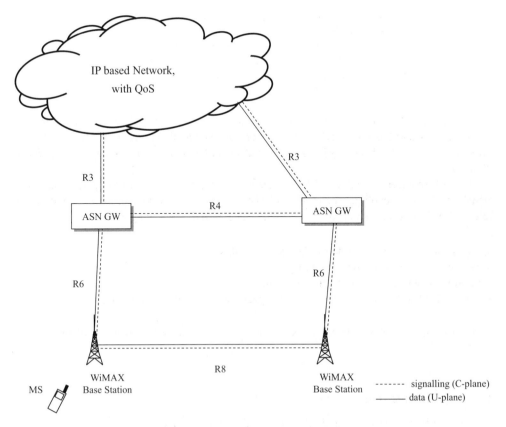

Figure 1.52 Mobile WiMAX network architecture.

IP-based core network. The ASN-GW is involved in the registration, mobility management, authentication, paging, billing and the service flow control.

General characteristics of the first IEEE 802.16e, also called mobile WiMAX release 1.0, are [11]:

- The duplexing is TDD.
- The physical layer adopts OFDMA multiple access, with QPSK, 16-QAM and 64-QAM modulations.
- The choice of the modulation and coding scheme is adapted to the radio channel conditions.
- The bandwidth is variable and can be 1.25, 5, 10 and 20 MHz.
- Subcarrier spacing is independent from the bandwidth, and is of 10.94 kHz.
- The MAC layer is connection oriented, based on the assignment of a group of subcarriers in a given time. It supports quality of service (QoS). This is valid for both the uplink and the downlink.
- Frequencies are licensed, in the range 3.4–3.6 GHz in most countries.
- Possibility of MIMO.

Table 1.17 Mobile WiMAX number of subcarriers for different
channel bandwidths

System BW (MHz)	1.25	5	10	20
Number of subcarriers	128	512	1024	2048

Because of the different characteristics, mobile WiMAX is not backward compatible with fixed WiMAX and has good performances even for fixed applications. Table 1.17 shows the different number of subcarriers in the case of channel bandwidths (BWs) of 1.25, 5, 10 and 20 MHz.

The performances of mobile WiMAX release 1.0 are comparable to the performances of HSPA. In 2009, an evolution of mobile WiMAX came out, also known as mobile WiMAX release 1.5. It introduced an evolved radio interface with the following goals:

- Extension of the TDD profile to FDD paired bands;
- Multicast/broadcast services enabling;
- Improvement of MAC layer efficiency; and
- New MIMO configurations, to improve capacity and cell edge throughput.

The performances of mobile WiMAX release 1.5 are comparable to the performances of HSPA+.

1.3.2.2 IEEE 802.16m

In 2011 the IEEE 802.16m standard, also called mobile WiMAX release 2.0, was approved. IEEE 802.16m satisfies the International Telecommunication Union (ITU) requirements for an International Mobile Telecommunications (IMT)-advanced (4G) system. It can operate in a licensed spectrum allocated to mobile or fixed broadband services, such as:

- 450–470 MHz
- 698–960 MHz, also identified for IEEE 802.16e
- 1710–2025 MHz
- 2110–2200 MHz
- 2300–2400 MHz, also identified for IEEE 802.16e
- 2500–2690 MHz, also identified for IEEE 802.16e
- 3400–3600 MHz, also identified for IEEE 802.16e

IEEE 802.11m is backward compatible with IEEE 802.11e, and adds some features to meet IMT-advanced requirements, such as:

- Carrier aggregation: two or more carriers, each with a bandwidth up to 20 MHz, can be aggregated. Contiguous and noncontiguous carriers can be aggregated.
- Extended MIMO configurations: up to an 8 × 8 MIMO configuration is considered in downlink; up to a 4 × 4 MIMO configuration is considered in uplink.

- Multibase station MIMO: includes with downlink coordinated transmission among different base stations. Improves high data rate coverage and cell-edge throughput.
- Relaying: used to improve the coverage of high data rates, cell-edge throughput, provide coverage in new areas.
- Enhanced multicast and broadcast services (E-MBS): offered on multicast connections to improve performance and operation in the power save mode.

1.3.3 Wireless PAN

A wireless personal area network (WPAN) is a wireless network with a very short coverage. There are many standards related to WPANs, with different properties like frequency, setup time, bit rates, number of nodes, etc. This section lists the most commonly used WPANs, with their main characteristics.

1.3.3.1 Radio Frequency Identification

A radio frequency identification system consists of two elements:

- The transponder, or tag, placed on the item to be identified
- The reader, or detector, able to read the tag

RFiD tags can be passive, if they are not alimented by a battery; active, if are alimented by a battery that is used to feed the radio communication part; semi-passive, if are equipped with a battery that feeds, for example, a memory or an onboard sensor but is not used for radio communication.

The coupling in passive RFiD can be:

- Inductive, based on the generation of a current induced by a magnetic field generated from the reader
- Backscatter, based on the transmission of an electromagnetic wave that is partially reflected and eventually modulated (modulated backscatter)

Frequency ranges for RFiD systems can be of low, medium or high frequencies, as shown in Table 1.18. RFiD uses different frequencies, different standards and different typologies of tags.

1.3.3.2 Near Field Communication

Near field communication (NFC) can be considered as an evolution of contactless RFiD technology. It can be integrated in a cell phone or in an SIM and can operate in three modes:

- As an emulation of a passive tag. In this case the cell phone can become a badge, a ticket, a credit card, etc.
- As a tag reader.
- To make a peer-to-peer connection with another device in order to exchange data, etc.

Table 1.18 Frequency ranges for RFiD, with standards and principal applications

Frequency	Standards	Application
125.5 kHz 134.2 kHz	ISO 14223 ISO 11785 ISO11784 ISO 18000-2	Animal identification
13.553–13.567 MHz	ISO 18092 ISO 14443 ISO 15693 ISO 18000-3	Contactless smartcards
433.05–434.79 MHz	ISO 1800-7	Remote control
868–870 MHz (EU) 902–928 MHz (USA) 960 MHz (Japan)	ISO 18000-6	Logistic and object identification
2.4–2.483 GHz	ISO 18000-4	Several systems, including vehicle identification (ISO 1800-4)

NFC works at the frequency of 13.56 MHz, has a range of a few centimetres, is based on ISO 18092 standard and is compatible with ISO 14443.

In March 2004 the NFC Forum was constituted, with the aim to ensure a minimum level of interoperability between devices and to develop specifications that are not included in the ISO standard. NFC can be considered as a service enabler, mainly for e-payments and m-payments.

1.3.3.3 Bluetooth and IEEE 802.15.1

Bluetooth was born as an industrial standard working in the 2.4 GHz Industrial, Scientific and Medical (ISM) band, promoted by Ericsson and then developed by the Special Interest Group (SIG), an association constituted in 1999 from Sony Ericsson, IBM, Intel, Toshiba, Nokia and other societies. The IEEE 802.15.1 standard was published in 2002 and is based on the Bluetooth v1.1 specifications.

Bluetooth v1.1 divides the spectrum into 1 MHz sub-bands, and uses frequency hopping as the multiple access technique. The modulation is Gaussian frequency shift keying. The Bluetooth range is of the order of magnitude of metres or tens of metres.

Bluetooth devices are able to create a network called a piconet. In the first phase, called *inquiry*, a wireless station acting as a master searches for other devices that are in the direct neighbourhood, called slaves. Then the *page* phase follows, where a slave device requests a connection. The master assigns physical resources, like hopping frequency and ciphering keys. The physical bit rate shared among the devices in a piconet is 1 Mbps in the Bluetooth v1.1. Concatenated piconets form a scatternet. In Figure 1.53, three piconets form a scatternet.

After Bluetooth v1.1, other versions were standardized. Bluetooth v1.2 includes some enhancements to Bluetooth v1.1, like faster connection and discovery, improvements of voice and audio quality. Bluetooth v2.1 is backward compatible with v1.2 and increases the maximum bit rate up to 3 Mbps. Bluetooth v.3 is for high bit rates, up to 24 Mbps. In this case the Bluetooth link is used for connection establishment and a WiFi link is used for high bit rate data transmission. Bluetooth v.4 is a protocol suite containing classic Bluetooth, Bluetooth

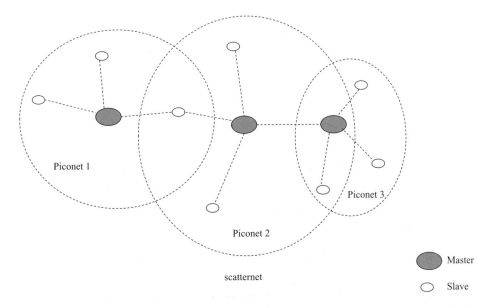

Figure 1.53 Bluetooth piconets and scatternet.

high speed (based on WiFi) and Bluetooth low energy. Bluetooth low energy includes a new protocol stack for low power consumption, low bit rate of about 250 kbps and short setup time.

Bluetooth low energy and near field communication (NFC) can both be integrated in mobile phones and can both be used as short range communication technologies.

1.3.3.4   IEEE 802.15.3

IEEE 802.15.3 is a standard, published in 2003, for high bit rate WPANs, which defines physical and MAC protocols for a network supporting up to 245 wireless devices in an area of tens of metres, with bit rates ranging from 11 Mbps to 55 Mbps. IEEE 802.15.3 uses 15 MHz wide channels in the 2.4 GHz unlicensed band. The choice of modulation and trellis code redundancy depends on the conditions of the radio channel.

The MAC protocol is based on a master–slave model in which a device, the piconet controller (PNC), controls the piconet. The PNC main tasks are:

• To manage device associations and disassociation,
• To coordinate the access to the radio channel of the devices forming the piconet,
• To coordinate the activities in the power save mode, and
• To manage the coexistence with other piconets sharing the same radio channel.

The piconet can support different types of traffic, like audio and video streams or best effort. The PNC sends a beacon that indicates the beginning of a superframe, consisting of two periods:

• Contention access period (CAP), based on CSMA and used for best effort traffic
• Channel time allocation period (CTAP), used for streaming or real time services

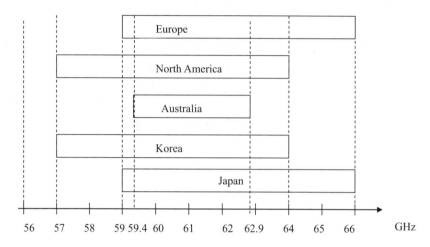

Figure 1.54   Available bandwidth in the 60 GHz band.

To transmit in the CTAP, a device requests time allocations to the PNC. If granted from the PNC, the device has exclusive access to the requested resource. A piconet can have one or more PNCs in the form of child PNCs. The PNC allocates to the child piconet coordinator a CTAP, where it transmits the beacon and allocates the resources to the devices belonging to the child piconet.

In 2009, a new IEEE standard for very high data rates, the 802.15.3c, was published. It consists of a new physical layer operating in the unlicensed band from 57 to 64 GHz, able to reach physical bit rates of up to 5.3 Gbps. Figure 1.54 shows the available bandwidths in Japan, Korea, Australia, Europe and North America. There is at least a 3.5 GHz contiguous spectrum available worldwide.

In North America, four channels of 2.16 GHz centred at 58.320 GHz, 60.480 GHz, 62.640 GHz and 64.800 GHz are defined. The radio is based on single carrier transmission with 14 different modulation and coding schemes, with bit rates up to 5.3 Gbps. The MAC layer is based on IEEE 802.15.3 MAC, with some enhancements. Typical applications for such bit rates are high definition uncompressed streaming video, interactive gaming, digital photography and digital home movies.

1.3.3.5   IEEE 802.15.4 and Zigbee

The IEEE 802.15.4 [12] standard defines the PHY and MAC for very low power, low bit rate network links. The radio transmission is based on the direct sequence spread spectrum (DSSS) on the following unlicensed bands:

- From 868 to 868.6 MHz, with a 600 KHz wide channel,
- From 902 to 928 MHz, with 10 channels of 2 MHz, and
- From 2.4 to 2.495 GHz, with 16 channels of 5 MHz.

The bit rate is of 20 kbps in the 868 MHz band, of 40 kbps in the 900 MHz band and of 250 kbps in the 2.4 GHz band. The modulation chosen for transmission in the bands of 868 and 900 MHz is BPSK; in the band of 2.4 GHz offset quadrature phase shift keying (OQPSK) is used.

The physical layer also defines procedures for:

- Activation and deactivation of the radio transceiver,
- Energy detection (ED): to select the appropriate channel,
- Link quality indication (LQI): to feedback the quality of a received packet, and
- Clear channel assessment (CCA): to determine if the channel is available for transmission.

The MAC protocol is based on the optional use of a superframe structure. There is a coordinator that defines the superframe structure, starting from a beacon frame and composed of 16 slots. The beacon is sent from the coordinator for synchronization purposes and contains network information and the superframe structure.

The superframe can be divided into two periods:

- Contention access period (CAP), based on CSMA
- Contention free period (CFP), where the coordinator allocates time slots to the devices that have requested guaranteed resources

The superframe can have inactive periods where devices turn to the power save mode. If the coordinator does not want to enable the superframe, it does not send the beacon and all transmissions follow CSMA/CA to access the channel.

Zigbee architecture is an enabler of sensor networks and Internet of things. A Zigbee network is formed of devices that can be of two types: full function device (FFD) or reduced function device (RFD). The coordinator must be a full function device.

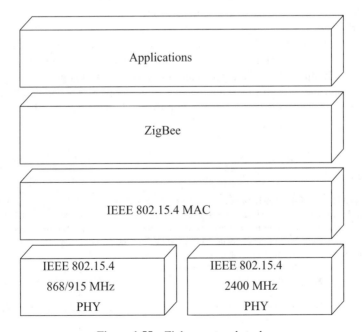

Figure 1.55 Zigbee protocol stack.

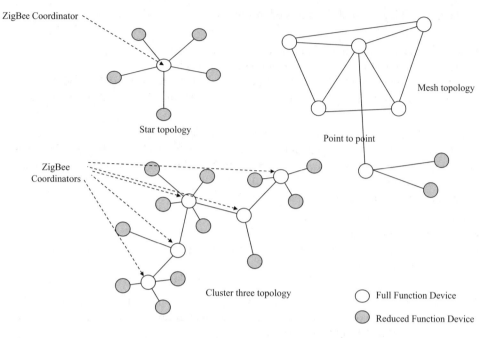

ZigBee Coordinator

Star topology

Mesh topology

Point to point

ZigBee
Coordinators

Cluster three topology

○ Full Function Device

● Reduced Function Device

Figure 1.56   Zigbee network architectures.

The Zigbee protocol stack is shown in Figure 1.55. It is based on IEEE 802.15.4 physical and MAC layers and defines the upper layer protocols to create a multihop ad hoc network. The Zigbee layer includes the following features: form a network, enter and leave a network, routing, security, etc.

Zigbee network topology can be star, peer-to-peer and cluster three. Network topologies are shown in Figure 1.56. Cluster three topology is obtained by combining star, point-to-point and mesh topologies. Reduced function devices are limited to star topologies and are able to talk only with the coordinator. Full function devices can be coordinators and can be used in any network topology. A Zigbee network can be of up to 65 534 nodes; a coordinator can manage up to 255 active nodes at a time.

References

1. Mouly, M. and Pautet, M.-B. (1992) *The GSM System for Mobile Communications*, Telecom Publishing.
2. Sanders, G., Thorens, L., Reisky, M. *et al.* (2003) *GPRS Networks*, John Wiley& Sons, Ltd, Chichester.
3. Eylert, B. (2006) *The Mobile Multimedia Business: Requirements and Solutions*, Halsted Press.
4. Kaaranen, H., Ahtiainen, A., Laitinen, L. *et al.* (2005) *UMTS Networks: Architecture, Mobility and Services*, 2nd edition, John Wiley & Sons, Ltd, Chichester.
5. A Regulatory Framework for UMTS, UMTS Forum Report 1, 1997.
6. Holma, H. and Toskala, A. (2009) *LTE for UMTS: OFDMA and SC-FDMA Based Radio Access*, John Wiley & Sons, Ltd, Chichester.
7. Alamouti, S.M. (1998) A simple transmit diversity technique for wireless communications. *IEEE Journal on Selected Areas in Communications*, **16** (8), October.

8. 3GPP TR 36.913 (2008) 3rd Generation Partnership Project; Technical Specification Group Radio Access Network; Requirements for Further Advancements for E-UTRA (LTE-Advanced), May 2008.

9. Pahlavan, K. and Levesque, A.H. (2005) *Wireless Information Networks*, John Wiley & Sons, Ltd, Chichester.

10. Mobile WiMAX – Part I: A Technical Overview and Performance Evaluation, prepared on behalf of the WiMAX Forum, 2006.

11. Mobile WiMAX – Part II: A Comparative Analysis, prepared on behalf of the WiMAX Forum, 2006.

12. Ergen, S.C. ZigBee/IEEE 802.15.4 Summary. http://www.eecs.berkeley.edu/csinem/academic/publications/zigbee.pdf.

2

Cognitive Radio: Concept and Capabilities

2.1 Cognitive Systems

A cognitive system is a complex adaptive system based on a sensing–understanding–action cycle [1]. An example of a cognitive cycle is the brain, consisting of a network of elements, the neurons, interacting with themselves and with their environment. The neurons sense the actual situation, understand it and act to reach a certain advantage or a given target. A cognitive system like the brain is able to adapt to changes in the environment and to react to them. The cognition cycle is shown in Figure 2.1.

Joseph Mitola in 1999 introduced the following definition for cognitive radio: 'the point in which wireless Personal Digital Assistants (PDAs) and the related networks are sufficiently computationally intelligent about radio resources and related computer-to-computer communications to detect user communications needs as a function of use context, and to provide radio resources and wireless services most appropriate to those needs' [2]. The radio cognition cycle is represented in Figure 2.2.

During the sensing phase, the cognitive device makes a real time wideband monitoring of the radio environment. Then, during the understanding phase, it makes a reliable characterization of the acquired context and chooses the best response strategy related to the context and user needs. Finally, during the action phase it changes radio parameters to provide the most appropriate radio resources and wireless services.

There are two recurrent definitions for cognitive radio:

- Spectrum sensing cognitive radio: only the radio frequency spectrum is considered. In this case the PDA senses the available spectrum, selects the best available channel, uses the selected channel in coordination with other users and releases it when a licensed user is detected [3].
- Full cognitive radio: every possible parameter observable by a wireless node or network is taken into account to optimize the whole. In this case the PDA acquires the context (i.e., available radio access technologies and operators, power and spectrum policies, actual utilization of the radio access technologies, or RATs, etc.) and, together with cognitive

Reconfigurable Radio Systems: Network Architectures and Standards, First Edition. Maria Stella Iacobucci.
© 2013 John Wiley & Sons, Ltd. Published 2013 by John Wiley & Sons, Ltd.

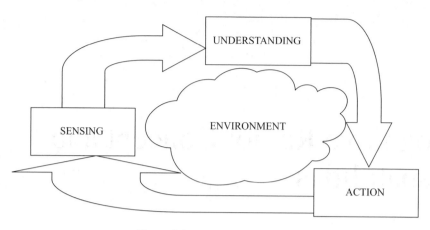

Figure 2.1   The cognition cycle.

network nodes, chooses the best radio access technology (RAT) and eventually reconfigures radio parameters to optimize radio resources utilization. Full cognitive radio systems lead to the reconfigurable radio systems, where each radio parameter, including the used spectrum, can be changed to meet user needs and network optimization purposes.

2.2   Spectrum Sensing Cognitive Radio

Wireless systems today can be divided in two classes:

- Licensed systems: the frequencies used are licensed and fixed, with limited spectrum allocation and optimization functions. In those systems (i.e., radiomobile systems like GSM/GPRS/EDGE, UMTS/HSPA/HSPA+, LTE) the interference is under control of the operator, but the usage of the spectrum is not optimized.

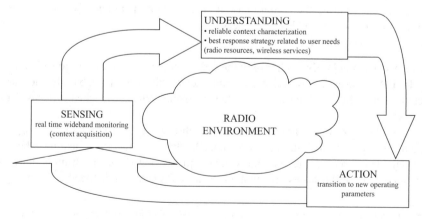

Figure 2.2   The radio cognition cycle.

Table 2.1   North American TV channel allocation

Channel number	Frequency band	
2, 3, 4	54–72 MHz	VHF
5, 6	76–88 MHz	
7, 8, 9, 10, 11, 12, 13	174–216 MHz	
14, 15, 16, 17, 18, 19, 20	470–512 MHz	UHF
21, 22, 23, ... , 51	512–698 MHz	

- Unlicensed systems: they work in unlicensed bands. Different systems use the same unlicensed bands (i.e., the 2.4 GHz band is used by WLANs, Bluetooth, Zigbee, etc.) without access coordination. This causes high interference and performance degradation.

Spectrum sensing cognitive radio systems provide dynamic spectrum management techniques that are able to use the available spectrum at that moment in that place by limiting the interference. Today the radio spectrum, although it is a scarce resource, is underutilized. The range of frequencies considered for spectrum sensing cognitive radio is the TV band. TV bands are represented in Tables 2.1 and 2.2 respectively for the United States and Europe.

The vacant TV channels are called white spaces and are considered for secondary spectrum use, which is the possibility for a cognitive radio device to transmit on the TV channels that are not used by TV repeaters (licensed primary users). Figure 2.3 shows that in United States there are many regions with up to 47 free TV channels.

In 2009, the first standard for cognitive radio, the IEEE 802.22, came out. IEEE 802.22 is addressed in Chapter 4 of this book.

A cognitive terminal sharing the radio spectrum with other users must implement new radio functionalities involving not only the physical but also the MAC and eventually network layers. The physical layer will be enhanced with:

- Sensing and geo-location functionalities, with the task of reliably measuring the radio frequencies in a certain geographic area,
- Cognitive functionalities, with the ability to process the spectrum measurements and choose the best available channels, and
- Adaptation functionalities, including the capability to move from an available frequency to another available frequency (frequency agility) and to adjust the power and the transmission format (i.e., modulation and coding) to the radio conditions.

Table 2.2   European TV channel allocation

Channel number	Frequency band	
5, 6, ... , 12	174–230 MHz	VHF
21, 22, ... , 69	470–862 MHz	UHF

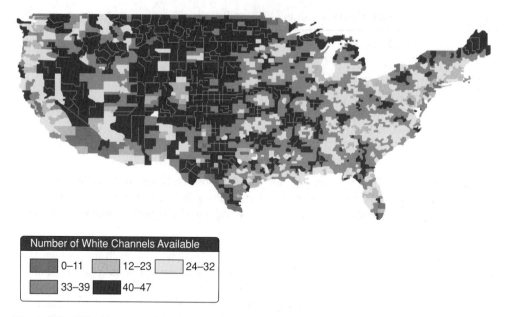

Figure 2.3 TV white space channels by county (December 2009). (Reproduced from www.dailywire less.com.)

The MAC layer will be enhanced with cooperation and spectrum management functionalities to enable the exchange of information among cognitive radio terminals in order to optimally share the available spectrum. If present, the spectrum renting functionality is also enabled at the MAC layer. Figure 2.4 shows the spectrum sensing cognitive features added to the MAC and physical layers.

2.2.1   Spectrum Sensing Cognitive Features

Figure 2.5 shows the sensing–understanding–action cycle for the cognitive features at the physical layer.

The sensing function includes wideband spectrum sensing and geo-location capabilities. The sensing results are used from the understanding function, which exploits the available radio frequencies and the transmission power and format (i.e., modulation and coding scheme) that best meet the user needs. Then, the action function selects the frequency, adjusts the power and adapts the transmission format. The described cycle continuously works together with the MAC to optimize spectrum usage while minimizing the interference and maintaining QoS/service requirements.

2.2.1.1   Wideband Spectrum Sensing

The functionality of wideband spectrum sensing is a very challenging research topic [4, 5]. Wideband spectrum sensing is a feature implemented at the physical layer.

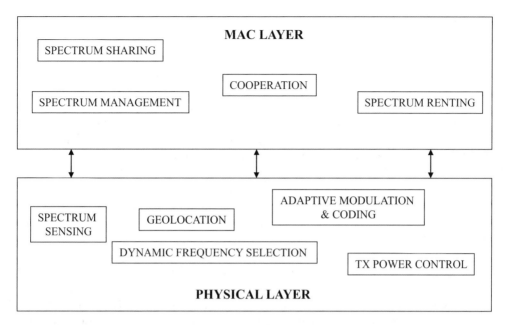

Figure 2.4   Spectrum sensing cognitive features added to the MAC and physical layers.

The first requirement for the spectrum sensing cognitive radio is to avoid interference to primary users. In order to achieve this goal, reliable detection of primary users is needed. At the receiver side, this is achieved by designing receivers able to reach stringent requirements on radio sensitivity. The radio sensitivity is the minimum signal power level required at the receiving antenna input to produce an output having a specified signal to noise ratio (SNR).

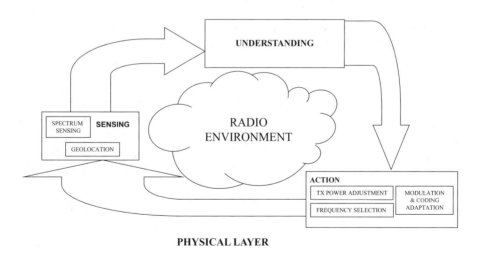

Figure 2.5   Sensing–understanding–action cycle for the cognitive features at the physical layer.

The formula for the receiver sensitivity S is:

$$S = \left(\frac{S}{N}\right)_{min} KT_0B\,(NF) \tag{2.1}$$

where K is Boltzmann's constant (38×10^{-23} J/EKelvin), T_0 is the absolute temperature (EKelvin) of the receiver input, B is the receiver bandwidth (Hz) and NF is the noise figure. The minimum power SM at the receiving antenna of a cognitive node must take into account the antenna gain G together with margins (M) and losses (L):

$$SM = \frac{S}{G}LM \tag{2.2}$$

which expressed in dB is:

$$SM_{dB} = S_{dB} - G_{dB} + L_{dB} + M_{dB} \tag{2.3}$$

In Equations (2.2) and (2.3), S represents the receiver sensitivity, G is the antenna gain, L includes losses and M takes into account all the margins (i.e., the set of parameters that must be taken into consideration to assure the required service quality). With respect to the traditional link budget problem, the new question is what signal power must be detected to avoid false alarms while guaranteeing no interference.

Figures 2.6 and 2.7 show two examples of a false alarm and a missed detection. In the example energy detection is performed. The false alarm implies that an available channel will not be used; the missed detection implies that interference to a primary user will be produced.

A cognitive radio reception chain is represented in Figure 2.8 [2]. The wideband RF signal incident on the antenna includes, other than noise, all the other interfering signals. The presence of strong and weak signals requires a large dynamic range for the front-end circuitry and in particular for the analogue to digital (A/D) converter, which must accommodate the large signals while still representing small signals in the range. Therefore, the wideband sensing requires high resolution A/D converters, with a dynamic range of up to 16 bits [6].

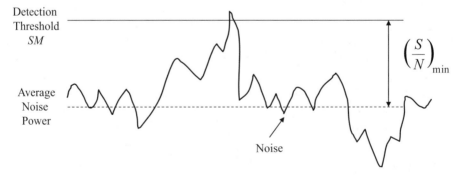

Figure 2.6   Example of a false alarm.

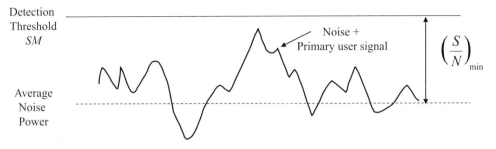

Figure 2.7   Example of a missed detection.

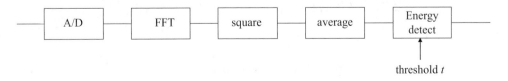

Figure 2.8   Example of a cognitive radio reception chain.

One possibility to reduce the range of the signal dynamic is to reduce the strong in-band primary users that are of no interest to detect, for example by filtering them with a notch filter [5]. Another possibility is to reduce the dynamic range of the received signal by filtering it in the spatial domain, for example with multiple antennas implementing beamforming.

Beamforming is an antenna technique that creates a directional radiation pattern, thus limiting the reception to the signals transmitted in that space/direction.

Detection Techniques
In this paragraph the four most important techniques used for signal detection in cognitive radio receivers are presented.

The first is based on **energy detection**, which is a quite simple concept. The detector measures the energy of the received signal and decides if a signal is present or not by threshold comparison. Because signal energy can be computed both in the time and frequency domains, the energy detector can be implemented in the time and frequency domains. Figure 2.9 shows a frequency domain implementation of an energy detector.

The frequency domain implementation of an energy detector works as follows: the received signal is digitalized and then converted to the frequency domain by performing a fast Fourier

Figure 2.9   Frequency domain implementation of an energy detector.

transform (FFT), squaring the coefficients and then averaging them. The obtained energy is compared to a threshold t to decide if the desired signal is present or not [7].

This is a noncoherent approach, which has the advantage of simple implementation also for wideband sensing. It does not require the knowledge of the signal to detect, but it suffers from low signal missed detection. In this case, an adaptive threshold setting can improve detection performances. Adaptive thresholding changes the threshold dynamically over the signal range. This more sophisticated version of thresholding can accommodate changes in the signal range without missed detection.

The second detection method is based on the **matched filter**, which leads to the optimum detection [4] of a known signal in random noise, with output signal to noise ratio (SNR) maximization. The impulse response of the matched filter is a delayed version of the mirror image of the input signal $s(t)$. Figure 2.10 shows a matched filter detector.

A matched filter can be realized through a correlation operation, between the received signal $r(t)$ and the impulse response of the filter $h(t)$:

$$y(t) = r(t) * h(t) = \int_{-\infty}^{\infty} r(\tau) h(t - \tau) d\tau \qquad (2.4)$$

or, in its discrete time version:

$$y_n = \sum_{k=-\infty}^{\infty} r_k h_{n-k} \qquad (2.5)$$

Matched filter detection applied to a cognitive radio means that the receiver knows the signal to be detected. Moreover, the coherent detection needs the receiver to perform timing and carrier synchronization. Different classes of primary users' signals can be stored in the matched filter detector, and in most cases synchronization is possible because most primary users broadcast pilots and synchronization signals. However, the matched filter detector presents a main drawback: the need to have a receiver for each primary user class, which makes its implementation very difficult [4]. An example of a cognitive radio detection based on a matched filter is shown in Figure 2.11. The detection is carried out by comparing the output of the matched filter with a threshold t.

The third detection method is **eigenvalue-based sensing**. The property that the statistical covariance of the signal and noise are different can be used to detect the presence of primary users in the received signal [8,9].

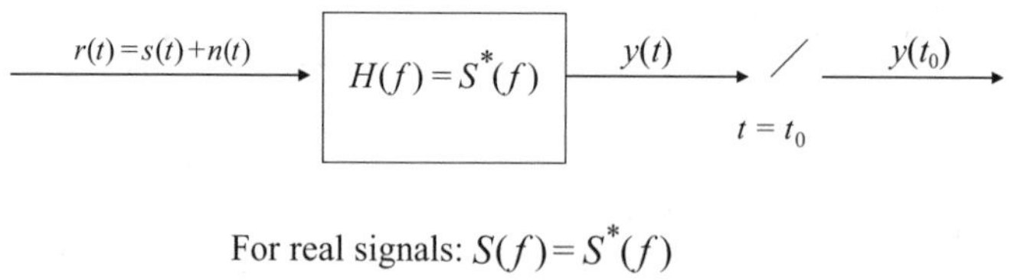

$$r(t) = s(t) + n(t) \longrightarrow \boxed{H(f) = S^*(f)} \xrightarrow{y(t)} \diagup \xrightarrow{y(t_0)}$$
$$t = t_0$$

For real signals: $S(f) = S^*(f)$

Figure 2.10 Matched filter detector.

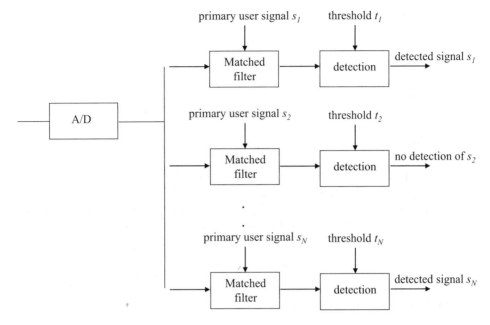

Figure 2.11   Example of cognitive radio detection based on a matched filter.

Let $r(t) = s(t) + n(t)$ be the received signal, which sampled is:

$$r(k) = s(k) + n(k) \qquad (2.6)$$

where $s(k)$ is the contribution of one or more primary user signals at the receiver side of the cognitive radio device and $n(k)$ is the sample of an additive Gaussian noise, with zero mean and covariance σ_n^2. Considering L consecutive samples, the following vectors can be defined:

$$\bar{r}(k) = [r(k)\, r(k-1) \dots r(k-L+1)] \qquad (2.7)$$

$$\bar{s}(k) = [s(k)\, s(k-1) \dots s(k-L+1)] \qquad (2.8)$$

$$\bar{n}(k) = [n(k)\, n(k-1) \dots n(k-L+1)] \qquad (2.9)$$

The statistical covariance matrices of $\bar{r}(k)$ and $\bar{s}(k)$ are respectively:

$$R_{\bar{r}} = E\left(\bar{r}(k)\,\bar{r}^{\mathrm{T}}(k)\right) \qquad (2.10)$$

$$R_{\bar{s}} = E\left(\bar{s}(k)\,\bar{s}^{\mathrm{T}}(k)\right) \qquad (2.11)$$

whose relation is:

$$R_{\bar{r}} = R_{\bar{s}} + \sigma_n^2 I_L \tag{2.12}$$

where I_L is an $L \times L$ identity matrix. The first consideration is that $R_{\bar{r}}$ is a diagonal matrix only in the case when the signal is absent and there is only white noise.

Many detection methods based on the computation of the received signal covariance matrix and eigenvalue calculation have been proposed [8–11]. Eigenvalue-based detection (EBD) methods can be divided into two classes:

- Semi-blind EBD, where an estimate of the noise covariance is available. This class provides better performances if the noise covariance is well known.
- Blind EBD, where the noise covariance is not known. Algorithms belonging to this class should be used in cases of unknown or highly variable noise covariance.

The different EBD algorithms compute a value T from a test statistic and compare it to a predetermined threshold t to decide if primary user signals are present or if there is only noise.

Suppose that H_0 is the event that only noise is present and H_1 is the event of existence of the primary user signal. The probability of false alarm is:

$$P_{fa} = \Pr(T > t \mid H_0) \tag{2.13}$$

and the probability of correct detection is:

$$P_{cd} = \Pr(T > t \mid H_1) \tag{2.14}$$

The missed detection probability is $P_{md} = 1 - P_{cd}$. The threshold t is usually set depending on a target false alarm probability. For example, in IEEE 802.22 P_{fa} and P_{md} must both be lower than 10^{-1}[11].

Let $\lambda_1, \lambda_2, \ldots, \lambda_L$ be the eigenvalues of $R_{\bar{r}}$. A semi-blind eigenvalue-based detection (EBD) algorithm consists of the computation of the maximum eigenvalue of $R_{\bar{r}}$, $\lambda_{\max}$, and defining $T = \lambda_{\max}/\sigma_n^2$ [11]. Primary user signals are detected if $\lambda_{\max} > t\sigma_n^2$. Observe that if $L = 1$ the maximum EBD algorithm turns to energy detection.

A blind EBD algorithm consists in the computation of T as follows [11]:

$$T = \frac{1}{L-1} \frac{\lambda_{\max}}{\sum\limits_{i=1}^{L-1} \lambda_i} \tag{2.15}$$

In Equation (2.15), the eigenvalues are sorted in ascending order and $\lambda_L = \lambda_{\max}$. The denominator of (2.15) is the maximum likelihood (ML) estimation of the noise variance if the presence of a signal is assumed [12]. Equation (2.15) substitutes for (2.14) in the case of unknown noise variance. Figure 2.12 shows the scheme of an eigenvalue-based detector.

In [13] it is shown that in the case of independent and identically distributed signals, energy detection is optimal. With correlated signals, EBD algorithms lead to better performances.

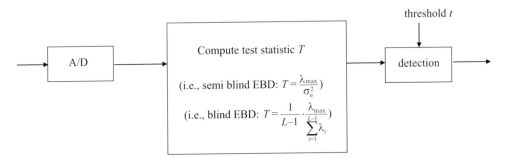

Figure 2.12 Eigenvalue-based detector.

Similar to energy detection, EBD does not need any prior knowledge about the signal to be detected [8, 10]. Moreover, no synchronization is needed.

The fourth method is the **cyclostationary detection** method. Primary users' signals include periodical signals like carriers, pulse trains and cyclic prefixes. The autocorrelation of such signals are characterized by a certain periodicity. The idea behind the cyclostationary detection method is to search for a periodicity in the autocorrelation of the received signal. If such periodicity is found, a primary user signal is detected.

The advantage of the cyclostationary detection method is that detection can be done based only on the received signal spectrum [14, 15]. The cyclic autocorrelation function (CAF) of the received signal $r(t)$ is defined as:

$$R_r^\alpha (\tau) = \lim_{T \to \infty} \frac{1}{T} \int_{-T/2}^{T/2} r\left(t - \frac{\tau}{2}\right) r\left(t + \frac{\tau}{2}\right) e^{-j2\pi\alpha t} \mathrm{d}t \qquad (2.16)$$

By calculating the Fourier transform (FT) of the CAF, the spectral correlation function (SCF) is obtained:

$$S_r^\alpha (f) = \int_{-\infty}^{\infty} R_r^\alpha (\tau) e^{-j2\pi\alpha\tau} \mathrm{d}\tau \qquad (2.17)$$

The SCF is a two-dimensional transform where the parameter α is the cyclic frequency. The spectral correlation function is also given by [3, 4]:

$$S_r^\alpha (f) = \lim_{T \to \infty} \lim_{Z \to \infty} \frac{1}{TZ} \int_{-Z/2}^{Z/2} R_T\left(t, f + \frac{\alpha}{2}\tau\right) R_T^*\left(t, f - \frac{\alpha}{2}\tau\right) \mathrm{d}t \qquad (2.18)$$

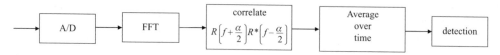

Figure 2.13 Cyclostationary detector.

where

$$R_T(t, f) = \int_{t-T/2}^{t+T/2} r(u)e^{-j2\pi f u}\, du \tag{2.19}$$

Equations (2.18) and (2.19) show that an implementation of a cyclostationary detector can be done by computing the fast Fourier transform (FFT) of the received signal $r(t)$, and then performing the spectral correlation. The digital implementation of the cyclostationary detector is shown in Figure 2.13. The detection is performed by searching the unique cyclic frequency corresponding to the maximum in the SCF plane. Cyclostationary detectors are difficult to implement, but they do not need prior knowledge of the signal to be detected and are more robust than energy detectors.

Cooperative Detection

Cooperation among secondary users can be adopted to improve primary user detection performances. The price you pay is the signalling among secondary users to exchange information about the detection of primary users [16]. On the other hand, thanks to the improved performances from cooperation, low complexity detectors like energy detectors can be used for cognitive devices [17].

Different cooperative detection schemes have been considered in literature. A cooperative detection scheme can be based on joint detection among different secondary users. For example, in [18] a fusion strategy is proposed of the decisions made by the individual nodes in the case for all cooperating nodes. The information about soft or hard decisions is exchanged through a control channel.

A different cooperative detection scheme implies that secondary users make independent decisions and communicate them to a fusion centre, which acts as a centralized controller performing the final decision and distributing it to the secondary users [19]. A different strategy is one where the future centre broadcasts the received information to the secondary users and each of them performs an individual decision taking into account the broadcasted information. Figure 2.14 shows the two described examples of cooperative detection.

2.2.1.2 Frequency Agility and Spectrum Aggregation for Dynamic Spectrum Access

After reliable detection of the available bandwidth (dynamic frequency selection), the cognitive radio terminal must be able to move to the available frequencies and use them efficiently. The dynamic and efficient usage of the available spectrum reduces spectrum fragmentation.

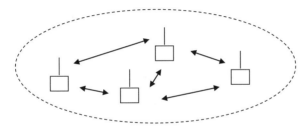

(a) Joint cooperative detection. The information is exchanged among the secondary users.

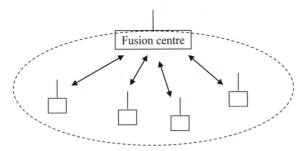

(b) Joint cooperative detection. The information is sent to a fusion center.

Figure 2.14   Examples of cooperative detection.

In this context, frequency-agile radios can improve the performance of wireless networks by dynamically occupying spectrum fragments. On the other hand, frequency agility implies a hardware complexity affecting the cost of the terminal. In [20] two types of frequency agile terminals are introduced:

- *1*-agile radios: are able to use one frequency channel but can adjust the channel width and central frequency
- *k*-agile radios: are able to use up to *k* noncontiguous frequency channels

The advantage of using *k*-agile radios is that the problem of spectrum fragmentation is solved at the physical layer. In the *1*-agile cognitive radio terminal, the output signal from the baseband module is converted to the central frequency of the fragment through the radio frequency (RF) transceiver and then transmitted to the front end module (FEM) connected to the antenna. Figure 2.15 shows a an *1*-agile radio transmission chain.

Figures 2.16 and 2.17 show two possible realizations of a *k*-agile cognitive radio terminal. A baseband module jointly elaborates the signals transmitted over each fragment and then a module aggregates the signals in transmission and disaggregates them in reception. In Figure 2.16, there is one RF module for each fragment. In Figure 2.17, a wideband RF module is used from all *k* fragments.

In the case of many fragments, it is impossible to have separate chains for each fragment and one chain must be shared among more than one fragment. This reduces the number of components but requires that each RF module be wideband. In that case, highly linear RF components designed for multicarrier applications should be used, thus increasing the cost of

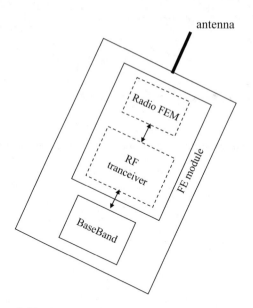

Figure 2.15   Example of an *1*-agile radio transmission chain.

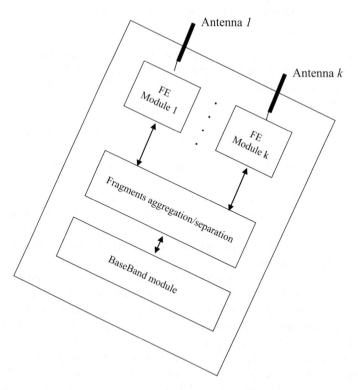

Figure 2.16   Example of a *k*-agile radio cognitive terminal with an RF module for each fragment.

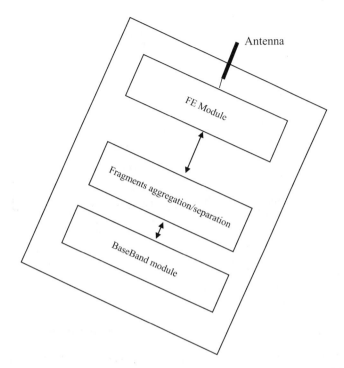

Figure 2.17 Example of a *k*-agile radio cognitive terminal with a wideband RF module shared among the *k* fragments.

the cognitive terminal [21]. Because of the size and coupling effects that can occur among close antennas, it would be appropriate to share one or two wideband antennas among different RF chains.

The transmissions schemes of frequency-agile radios described above solve the spectrum fragmentation problem at the physical layer. The problem of spectrum fragmentation can also be solved at higher layers, through the function of spectrum aggregation in the scheduler. Different spectrum aggregation algorithms for dynamic spectrum access have been proposed [21–24].

The aggregation-aware spectrum assignment (AASA) algorithm is a very simple algorithm assuming that all users have the same bandwidth requirement. It is based on the assignment of spectrum fragments starting from low frequencies [22] and its flow chart is shown in Figure 2.18. Because of its simplicity, the AASA algorithm can be implemented in both centralized and distributed networks [22, 23].

A different spectrum aggregation algorithm assuming that the users have different spectrum requirements is the maximum satisfaction algorithm (MSA). The idea is to serve first the user with the largest bandwidth requirement, to which will be assigned the fragment with the least fitting available bandwidth, and the other fragments will then remain available to the other users. The flow chart of the MSA is shown in Figure 2.19. The MSA is better suited for implementation in centralized networks.

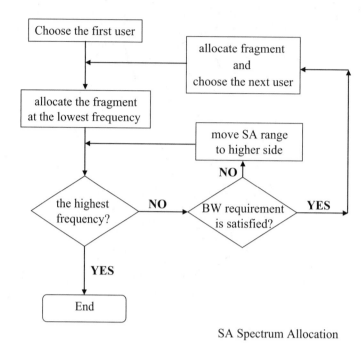

SA Spectrum Allocation

Figure 2.18 Aggregation-aware spectrum assignment algorithm flow chart.

Of course the AASA algorithm and MSA is not an exhaustive list of the available spectrum assignment algorithms but represent two examples of the fragment aggregation problem solved at the MAC layer.

2.2.1.3 Spectrum Sharing and Spectrum Renting

Spectrum sharing implies that the spectrum is shared among different subjects. The subjects share the available frequencies, which can be both licensed and unlicensed; the subjects sharing the spectrum can be base stations (BSs) or terminals. An example of spectrum sharing is shown in Figure 2.20, where two base station nodes share the same frequencies. The BSs can belong to the same or to different operators. Spectrum sharing also brings the opportunity of spectrum renting, which is the possibility of opportunistically offering the unused spectrum for renting. In Figure 2.20 at time t_1 BS1 uses a larger spectrum that BS2 because it needs to serve more users; at time t_2 two of the five users of BS1 move to the BS2 coverage. Then the used spectrum is rearranged between BS1 and BS2 to serve the new traffic load.

The other example is spectrum sharing among unlicensed users within the same band. The AASA algorithm and MSA described in Section 2.2.1.2 can be used for spectrum sharing among unlicensed users. In [25] and [26] the spectrum sharing problem among unlicensed users within the same band has been studied and fair and efficient spectrum sharing rules coming from the game theory have been proposed.

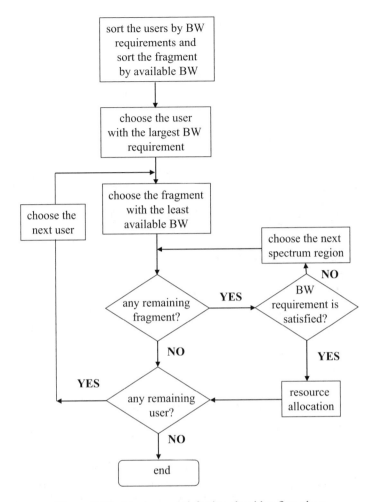

Figure 2.19 Maximum satisfaction algorithm flow chart.

2.2.1.4 Geo-location

Cognitive radio uses unoccupied bands for data transmission. Because the unused bands vary with position, the location information can be used for bandwidth allocation. In this method, a device determines its location and sends the location with its accuracy to a database. The database, which knows the available channels at that location, assigns one or more channels to the cognitive radio together with the maximum allowed power and time validity. At the same time, the cognitive device can send to the database its sensing results in order to update its information. This scenario with examples is described in Chapter 4 of this book.

   In this section, a short review of geo-location techniques is provided [27]. Geo-location techniques can be satellite based or terrestrial based. In the first case, the anchor points are satellites; in the second case, the anchor points are terrestrial nodes (i.e., base stations).

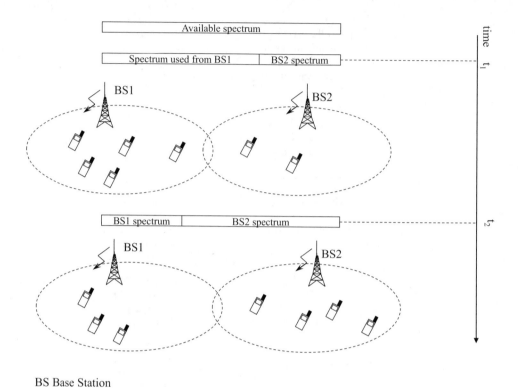

BS Base Station

Figure 2.20   Example of spectrum sharing between two base stations.

Different location techniques can be used in cognitive radio networks. The two fundamental principles used in traditional localization techniques are triangulation and trilateration.

With triangulation, the location is obtained by measuring the angles from known points. The point whose location must be determined is a vertex of a triangle with one known side and two known angles. In Figure 2.21 an example of triangulation is shown, where the known side

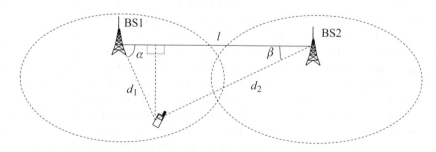

Figure 2.21   Example of triangulation.

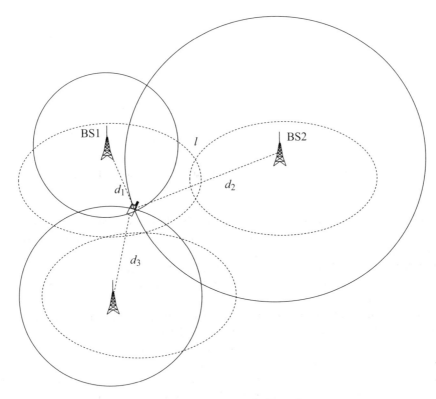

Figure 2.22 Example of trilateration.

is the distance between two base stations and α and β are the two known angles. The distance d_1 between the mobile terminal and BS1 is given by:

$$d_1 = \frac{l \sin(\beta)}{\sin(\alpha + \beta)} \tag{2.20}$$

The distance d_2 between the mobile terminal and BS2 is given by:

$$d_2 = \frac{l \sin(\alpha)}{\sin(\alpha + \beta)} \tag{2.21}$$

With trilateration, the location is obtained by measuring the distances from at least three points having a well-known position (anchor nodes). In Figure 2.22 an example of trilateration where the three anchors are three base stations is shown. The position is determined through the intersection of three circles of radii d_1, d_2 and d_3.

In practical implementations, the most popular methods are:

- **Cell identifier (cell-Id)**: is a network-based method where the location is determined by reading the cell broadcast information. The cell-Id identifies a cell, which is the region within

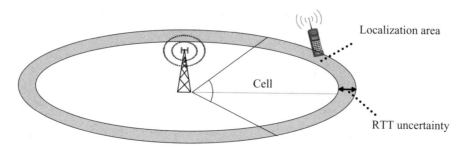

Figure 2.23   Cell-Id + round trip time localization method.

which the mobile terminal location is identified. Cell-Id is a very simple method but with the approximation of the cell dimension, which can range from a few tens of square metres to square kilometers.

- **Round trip time (RTT)**: the method calculates the RTT to determine the distance between a mobile terminal and an anchor node (i.e. a BS). The use of RTT in combination with cell-ID is shown in Figure 2.23.
- **Time of arrival (TOA)**: compares the time a signal transmitted from the mobile station arrives at three or more different anchor nodes to determine a two-dimensional position. Fundamentally, it is a trilateration technique. Variants of time of arrival are time difference of arrival (TDOA) and observed time difference of arrival (OTDOA).
- **Received signal strength (RSS)**: estimates the distance between the mobile and three or more anchor nodes through the measurement of the received strength of a known signal. It is a trilateration technique.
- **Angle of arrival (AOA)**: is used in triangulation because it estimates the direction of propagation of a radio wave incident on an antenna.

Of course, the anchor nodes can be base stations or, in an ad hoc network, other terminals with known position.

2.2.1.5   Adaptive Modulation and Coding

Spectrum sensing acts at the receiver side in order to detect the presence of primary users and choose the free spectrum channels for transmission. At the transmitter side, modulation and coding schemes and power levels are adjusted to avoid interference to primary users.

Modulation

In this subsection QAMs are revisited with the goal to provide a basic understanding of the involved parameters and the adaptation possibilities [28]. The transmission chain with modulation is shown in Figure 2.24.

The source emits $f_b = 1/T_b$ bits per second, which are modulated to be sent on the radio channel. The modulator emits f_s symbols per second. The modulated signal occupies the band B around the carrier f_c. The noise is additive and Gaussian with power spectral density n_0. The

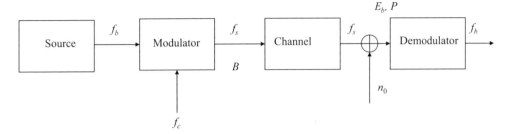

Figure 2.24   Transmission chain with modulation.

bandwidth B occupied from a modulated signal is directly related to the symbol frequency f_s. The minimum bandwidth required for a linearly modulated signal is:

$$B_{\min} = 2 f_N = f_s \tag{2.22}$$

where f_N is the Nyquist frequency. The power P of the received signal is related to the energy of a bit E_b from the following relation:

$$E_b = PT_b \tag{2.23}$$

The signal to noise ratio S/N is:

$$\frac{S}{N} = \frac{E_b f_b}{n_0 B} \tag{2.24}$$

The Shannon equation about the channel capacity [29]:

$$C = B \log_2 \left(1 + \frac{S}{N} \right) \tag{2.25}$$

highlights the relationship between the bandwidth, the signal to noise ratio and the maximum achievable capacity C. The Shannon theorem also establishes the existence of a channel coding that allows transmission without errors on a channel of band B at rate $f_b \le C$. Considering $f_b = C$ and then $S = E_b / T_b = E_b f_b = E_b C$ and $N = n_0 B$, the Shannon equation can be rewritten as follows:

$$\frac{C}{B} = \log_2 \left(1 + \frac{E_b}{n_0} \frac{C}{B} \right) \qquad \text{bit/sHz} \tag{2.26}$$

The Shannon curve represented in Figure 2.25 establishes a limit and a reference for a comparison among different transmission techniques.

The polar representation of a signal describes an oscillation through a phasor of a certain amplitude and phase. Figure 2.26(a) shows the phasorial representation of the signal:

$$s(t) = A \cos(\omega_0 t + \varphi) \quad \text{where } \omega_0 = 2\pi f_0 \tag{2.27}$$

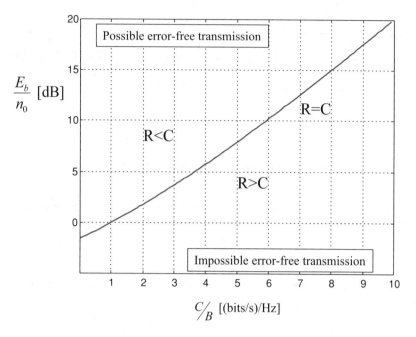

Figure 2.25 Shannon curve.

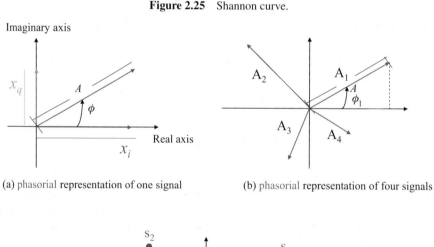

(a) phasorial representation of one signal   (b) phasorial representation of four signals

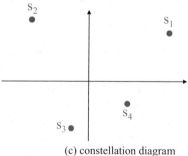

(c) constellation diagram

Figure 2.26 (a) Phasorial representation of one signal; (b) phasorial representation of four signals; (c) constellation diagram.

Considering the two orthogonal phasors x_i and x_q represented in Figure 2.26(a), the signal $s(t)$ can also be written as follows:

$$s(t) = x_i \cos(\omega_0 t) + x_q \sin(\omega_0 t) \qquad (2.28)$$

Figures 2.26(b) and (c) respectively represent the phasorial representation of four signals and the constellation diagram, obtained with the end points of the phasors.

The noise probability density function $p(n)$ is Gaussian with zero mean and variance σ^2:

$$p(n) = \frac{1}{\sqrt{2\pi}\sigma} e^{-x^2/2\sigma^2} \qquad (2.29)$$

The Q-function is:

$$Q(x) = \frac{1}{\sqrt{2\pi}} \int_x^\infty e^{-\lambda^2/2} d\lambda \qquad (2.30)$$

A simplified method for error probability estimation is now introduced for binary amplitude modulation and is then extended for QAMs. In Figure 2.27 is represented a 2PSK (phase

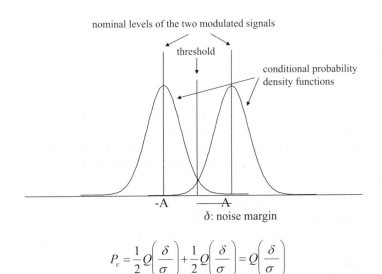

nominal levels of the two modulated signals

threshold

conditional probability density functions

-A $\qquad$ A

δ: noise margin

$$P_e = \frac{1}{2} Q\left(\frac{\delta}{\sigma}\right) + \frac{1}{2} Q\left(\frac{\delta}{\sigma}\right) = Q\left(\frac{\delta}{\sigma}\right)$$

Figure 2.27 Representation of a 2PSK modulation with amplitude A and additive Gaussian noise with zero mean and variance σ^2.

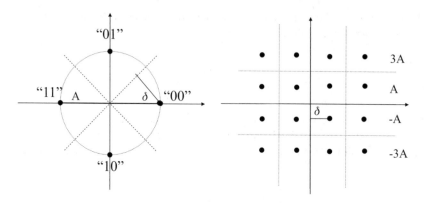

Figure 2.28   4-QAM and 16-QAM constellation diagrams.

shift keying) modulation with amplitude A and additive Gaussian noise with zero mean and variance σ^2. In that case, the error probability is given by:

$$P_e = \frac{1}{2}Q\left(\frac{\delta}{\sigma}\right) + \frac{1}{2}Q\left(\frac{\delta}{\sigma}\right) = Q\left(\frac{\delta}{\sigma}\right) = Q\left(\sqrt{\frac{\delta^2}{\sigma^2}}\right) \tag{2.31}$$

where δ is the noise margin and $\sigma^2 = n_0 B$ the noise variance. For the 2PSK:

$$S = \frac{A^2}{2} = \frac{\delta^2}{2} = E_b f_b \tag{2.32}$$

and then:

$$P_e = Q\left(\sqrt{2\frac{E_b}{n_0}}\right) \tag{2.33}$$

The proposed method is now extended to calculate the error probability of 4-QAM and 16-QAM. Then the exact error probability functions are given. Figure 2.28 shows 4-QAM and 16-QAM constellation diagrams with noise margin δ.

Table 2.3 reports the principal parameters of 4-QAM, 16-QAM and 64-QAM modulations under the assumptions described in this subsection. Also, the 64-QAM is supposed to have noise margin δ.

Table 2.3   4-QAM, 16-QAM and 64-QAM parameters

	4-QAM	16-QAM	64-QAM
Bandwidth occupation	$B = 2f_N = f_s = f_b/2$	$B = 2f_N = f_s = f_b/4$	$B = 2f_N = f_s = f_b/6$
Noise margin	$\delta = A/\sqrt{2}$	$\delta = A$	$\delta = A$
Average transmitted power	$S = A^2/2 = \delta^2$	$S = 5A^2 = 5\delta^2$	$S = 21A^2 = 21\delta^2$
Noise power	$n_0 B = n_0 f_b/2$	$n_0 B = n_0 f_b/4$	$n_0 B = n_0 f_b/4$

In QAM modulations we can approximate the error probability P_e with:

$$P_e \cong Q\left(\sqrt{\frac{\delta^2}{\sigma^2}}\right) \tag{2.34}$$

From Table 2.3 we can calculate that:

- For 4-QAM: $\delta^2 = S = E_b f_b, \sigma^2 = \dfrac{n_0 f_b}{2}$ and $P_e \cong Q\left(\sqrt{\dfrac{2E_b}{n_0}}\right)$
- For 16-QAM: $\delta^2 = \dfrac{S}{5} = \dfrac{E_b f_b}{5}, \sigma^2 = \dfrac{n_0 f_b}{4}$ and $P_e \cong Q\left(\sqrt{\dfrac{4E_b}{5n_0}}\right)$
- For 64-QAM: $\delta^2 = \dfrac{S}{21} = \dfrac{E_b f_b}{21}, \sigma^2 = \dfrac{n_0 f_b}{6}$ and $P_e \cong Q\left(\sqrt{\dfrac{2E_b}{7n_0}}\right)$

The proposed method for error probability derivation of QAMs is not rigorous but gives a good understanding of modulation parameters and their relations. Fixing the occupied bandwidth B, if the modulation cardinality (number of modulation symbols) increases the spectral efficiency also increases, but a higher signal to noise ratio is required to obtain the same error probability. Figure 2.29 shows the 4-QAM, 16-QAM and 64-QAM positions with respect to the Shannon curve for $P_e = 10^{-6}$.

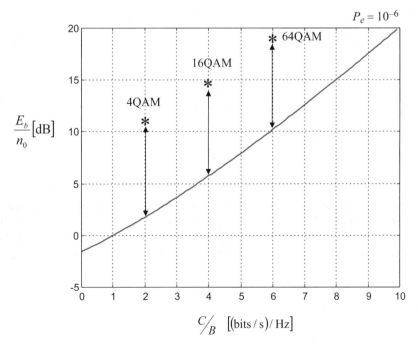

Figure 2.29 4-QAM, 16-QAM and 64-QAM positions with respect to the Shannon curve for $P_e = 10^{-6}$.

The exact error probability for QAMs is given by:

$$P_e = \left(1 - \frac{1}{L}\right) 4Q\left(\sqrt{\frac{2E_b}{n_0} \frac{3 \log_2 L}{L^2 - 1}}\right) \tag{2.35}$$

where L^2 is the number of QAM signals. If C is the Shannon capacity and f_b is the bit rate of the transmission system referred to in Figure 2.14, defined as the signal to noise ratio *gap* or simply gap, the quantity:

$$\gamma = \frac{2^{2C} - 1}{2^{2f_b} - 1} \tag{2.36}$$

The real bit rate f_b is then:

$$f_b = B \log_2\left(1 + \frac{S/N}{\gamma}\right) \tag{2.37}$$

The gap γ is a function of the chosen modulation and of the error probability P_e.

Coding
In this subsection the principles of error correcting coding are revisited with the goal to provide a basic understanding of the involved parameters and the adaptation possibilities. Figure 2.30 shows the transmission chain with modulation and coding.

The binary source emits f_b bits per second. The bits are encoded at code rate k/n. At the output of the encoder, the bit rate increases at $f_b' = (n/k) f_b$. Then, the channel coding requires a higher bandwidth than an uncoded transmission. Moreover, the energy associated to a bit at the output of the encoder is $E_b' = (k/n) E_b$.

The advantage of using error correcting codes is that a *coding gain* is obtained with respect to an uncoded transmission, as shown in Figure 2.31. The *coding gain* is the gain in SNR obtained for a given error probability P_e by using a certain error correcting code with respect to an uncoded transmission. In Figure 2.31, the 'uncoded' curve is referred to a 4-QAM, while the 'coded' curve is referred to the same modulation and a BCH error correcting code with $k/n = 15/31$. In that case, the coding gain for $P_e = 10^{-6}$ is $cg \cong 3$ dB. Observe that in the case shown in Figure 2.30 encoding gives a gain only for $E_b/n_0 \geq 4$ dB.

Figure 2.32 shows the curves of an uncoded 4-QAM transmission and the same transmission coded with convolutional codes having coding rates of 1/2, 1/3, and 1/4 . For the same code used, that is convolutional codes with the same constraint length, the higher the coding rate $R_c = k/n$, the higher is the coding gain for the same error probability P_e. The constraint length represents the number of bits in the encoder memory that affect the generation of n output bits.

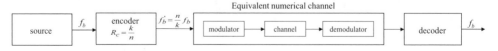

Figure 2.30   Transmission chain with modulation and coding.

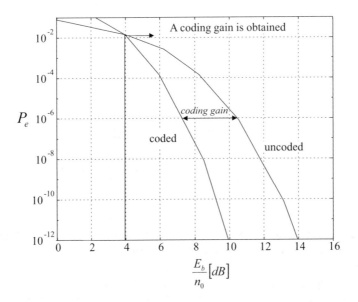

Figure 2.31   Example of P_e curves for coded and uncoded transmission.

Because the choice of transmission scheme (modulator and encoder) depends on the signal to noise ratio at the receiver side, a reporting procedure is needed in order to inform the transmitter on the channel status.

Figure 2.33 shows an example of a channel feedback reporting procedure in the case of downlink data transmission. The access network node notifies the mobile station with the

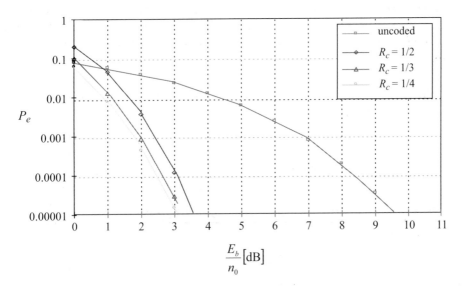

Figure 2.32   Example of P_e curves for convolutional codes with constraint length $= 9$ and different coding rates.

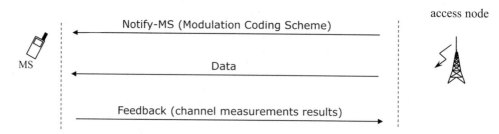

Figure 2.33 Example of channel feedback reporting procedure in the case of downlink data transmission.

modulation and coding scheme (MCS) what must be use for data reception; after the data reception the MS sends a feedback with channel measurements results, which will be used for the computation of the MCS in the subsequent transmission.

In Figure 2.34, an uplink transmission is shown. In that case the channel is measured from the access node, which assigns the MCS to the MS. The procedure that adapts the modulation and coding scheme to the SNR measured at the receiver is called adaptive modulation and coding (AMC).

Hybrid Automatic Repeat Request

An automatic repeat request (ARQ) is a function of the radio link control (RLC) layer, which uses an error-detecting code to verify that an RLC-PDU (RLC-packet data unit) has been correctly received. A positive acknowledgement is sent for a correct reception; a negative acknowledgement is sent for error detection. In the latter case the packet is retransmitted.

Hybrid ARQ (HARQ) is a technique that adaptively combines an error correcting coding with a cyclic redundancy check (CRC), used for error detection. There are three types of HARQ:

- Type I HARQ: in case of error detection the RLC-PDU is discarded and the retransmission is requested. It is the same as ARQ.

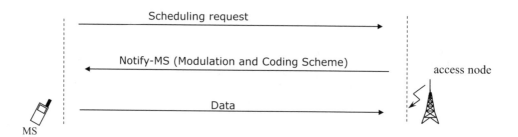

Figure 2.34 Example of uplink data transmission.

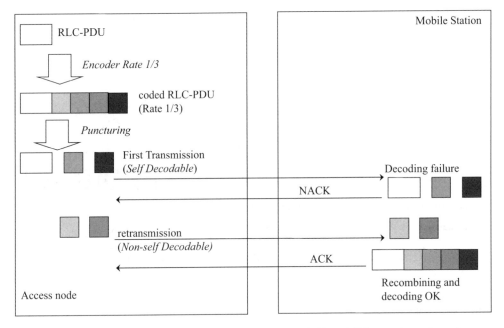

Figure 2.35 Example of type II HARQ with coding rate $R_c = {}^1\!/_3$ and different puncturing schemes.

- Type II HARQ: is also known as ARQ incremental redundancy (IR). In this case, an RLC-PDU received with errors is not discarded but is combined with incremental redundancy sent in subsequent retransmissions, which are not self-decodable.
- Type III HARQ: it differs from the previous case because it uses self-decodable retransmissions. In this case, an RLC-PDU received with errors is not discarded and can be retransmitted with the same or a different set of coded bits. The receiver operates soft combining of the received RLC-PDUs. If the RLC-PDU has been retransmitted with the same coded bits as the first transmission, then the combining is called chase combining. If the RLC-PDU has been retransmitted with different coded bits, this kind of type III HARQ belongs to IR HARQ. Incremental redundancy achieves better performances than chase combining, but requires more buffer memory at the receiver.

Figure 2.35 exemplifies type II HARQ in the case of coding rate $R_c = {}^1\!/_3$ and different puncturing schemes. With puncturing some of the redundancy bits after the encoding are removed, thus producing an encoded transmission with less redundancy.

Figure 2.36 shows an example of type III IR HARQ where the redundancy is incremented with the retransmissions. In fact, in the first transmission only the information bits (C1) are sent and the coding rate R_c is equal to one. Because the error detecting code (i.e. CRC) reveals some errors, the RLC-PDU is memorized and a retransmission is requested. In the second transmission the RLC-PDU is sent with coding rate $R_c = {}^1\!/_2$ (C1 + C2 are transmitted). Also, in the second transmission the error detecting code reveals some errors. The RLC-PDU is memorized and a retransmission is requested. In the third transmission the RLC-PDU is sent

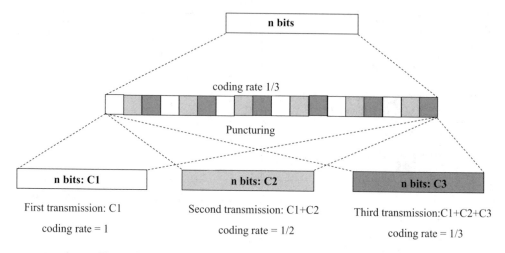

Figure 2.36 Example of type III incremental redundancy HARQ.

with coding rate $R_c = {}^1/_3$ (C1 + C2 + C3 are transmitted). All the received information is combined to obtain the correct information bits.

Cooperative HARQ

Cooperative communication uses information heard from neighbouring nodes to provide robust communication between two network elements (mobile terminals, access nodes) [30–33]. An example of cooperative communication with three nodes is shown in Figure 2.37. It concerns a source S, a destination D and a relay node R.

Cooperation can be implemented at the physical layer to create a transmit diversity and improve the SNR at the reception side. For example, the relay R could amplify and forward the signal transmitted from the source S. The destination will receive two signals carrying the same information, which experience different channels. Cooperation can also be implemented

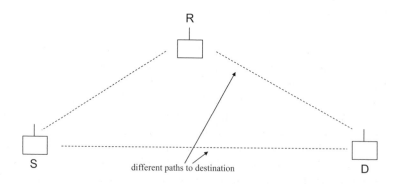

Figure 2.37 Example of a cooperative network with relay.

at the radio link control (RLC) layer through a cooperative HARQ scheme, or at both PHY and RLC (radio link control) through AMC and HARQ.

An example of cooperative HARQ is the following: the source sends an RLC-PDU, which contains a CRC. The destination, after CRC computation sends an ACK for correct reception or a NACK for incorrect reception. If a NACK is received from the relay, and has correctly received the packet, it retransmits the packet to the destination using HARQ techniques (i.e., only an incremental redundancy or the packet differently encoded is transmitted). The destination recombines the subsequent retransmissions, which can or cannot be self-decodable.

2.2.1.6   Transmit Power Control

Spectrum sensing acts at the receiver side in order to detect the presence of primary users and choose the free spectrum channels for transmission. At the transmitter side, other than AMC and HARQ, power levels are adjusted to avoid interference to primary users [34–36].

The considered scenario is a spectrum sensing cognitive radio system where a secondary radio transmitter detects the occupancy of frequency bands from primary users and communicates using the spectrum holes without causing interference to the primary user. The described scenario is represented in Figure 2.38.

The secondary transmitters vary the transmission power based on all the information sensed from the environment. A power control rule must be applied to guarantee the needed quality of service (QoS) to the primary system. Moreover, the maximum power allowed to a secondary user can be limited depending on regulation rules and/or its location. Suppose that a secondary user can transmit at maximum power P_{max}.

A very simple approach is based on binary detection, where the secondary user senses the spectrum and decides if in a certain frequency band a primary user is present or not. If a primary user is detected, the secondary user does not transmit; otherwise it transmits at the maximum power P_{max}. The success of this approach only depends on the probability of successful detection.

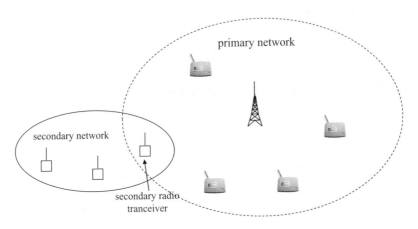

Figure 2.38   Scenario with primary and secondary networks.

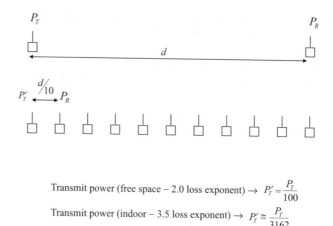

$$\text{Transmit power (free space} - 2.0 \text{ loss exponent)} \rightarrow P_T^r = \frac{P_T}{100}$$

$$\text{Transmit power (indoor} - 3.5 \text{ loss exponent)} \rightarrow P_T^r \cong \frac{P_T}{3162}$$

Figure 2.39 Example of a transmission power comparison in the case of a direct link and 10 relays.

A different approach provides a power adaptation scheme where the secondary users can vary the transmission power from zero to P_{max} as a function of the sensed information, with the goal of minimizing the interference to the primary users and maximizing the performances of the secondary users. In [35] it is shown that the power control based on binary detection leads to optimum secondary user's SNR performances.

On the other hand, the power adaptation scheme that optimizes the secondary user capacity, with maximum transmission power and average interference constraints at the primary receiver, is a continuous adaptive power control ranging from 0 to P_{max} and depending on the sensing metric (e.g., it can be the total observed power or the correlation between the observed signal and a known sequence, etc.) [35].

In the case of cooperative communications, the use of relays implies lower power levels and lower interference to primary users. Figure 2.39 shows a very simple example showing a direct link transmission and a transmission with 10 relays. The distance between the transmitter and the receiver is d for the direct link and is d/N in the case of N relays. The transmitted power is P_T in the case of a direct link and P_T^r in the case of N relays. We assume that the power P_R is received in both cases. In the case of free space and direct link, the received power is:

$$P_R = \alpha P_T d^{-2} \tag{2.38}$$

In the case of free space and N relays, the received power is:

$$P_R = \alpha P_T \left(\frac{d}{N}\right)^{-2} \tag{2.39}$$

Then the transmission power needed to receive the same power P_R is N^2 times lower with N relays than with direct link.

In the case of loss exponent 3.5 (i.e., indoor loss), the received power is:

$$P_R = \alpha P_T d^{-3.5} \tag{2.40}$$

In the case of free space and N relays, the received power is:

$$P_R = \alpha P_T \left(\frac{d}{N}\right)^{-3.5} \tag{2.41}$$

Then the transmission power needed to receive the same power P_R is $N^{3.5}$ times lower with N relays than with direct link. Figure 2.39 shows the particular case with 10 relays.

2.3 Introduction to the Full Cognitive Radio

With spectrum sensing cognitive radio terminals and networks, spectrum usage can be optimized among different radio access technologies. In this context, new scenarios with different degrees of freedom can be considered: from the case where unlicensed cognitive radio terminals operate in times and zones where a licensed spectrum is underutilized, to the scenario of a licensed operator that uses cognitive radio inside its network to increase the efficient use of radio resources, through the coordination of licensed operators for spectrum usage.

Cognitive radio can also open a scenario where the spectrum resource could be managed in an hour-to-hour market for spectrum exchange. In a full cognitive radio every possible parameter observable by a wireless node or network is taken into account for adaptation and network optimization. A full cognitive radio leads to reconfigurable radio systems in both network and terminal sides. The considered scenario is with heterogeneous access networks connected with a unique packet-based core network, as represented in Figure 2.40.

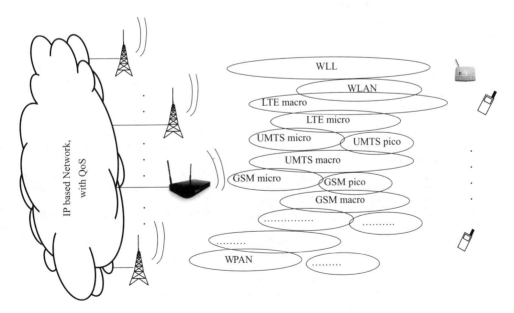

Figure 2.40 Heterogeneous access network scenario.

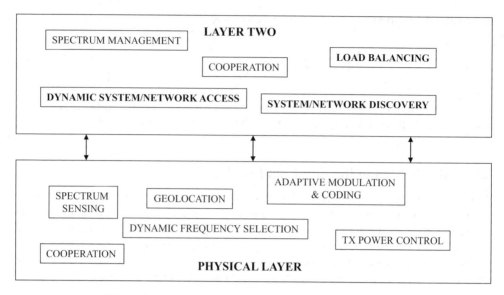

Figure 2.41   Another heterogeneous access network scenario.

In this scenario, the radio terminal implements all or part of the available RATs and chooses, autonomously or together with the network, the best RAT to be connected within the function of the service and/or network usage optimization. In order to reach this goal, new higher layer algorithms, like dynamic system/network access, system/network discovery and load balancing, must be implemented at the MAC layer, as shown in Figure 2.41. Examples of such algorithms are provided in other chapters in the present book.

References

1. Schuster, H.G. (2001) *Complex Adaptive Systems: An Introduction*, Scator Verlag.
2. Mitola III, J. and Maguire Jr, G.Q. (1999) Cognitive radio: making software radios more personal. *IEEE Personal Communications*, **6** (4), August, 13–18.
3. Chen, K.-C. and Prasad, R. (2009) *Cognitive Radio Networks*, John Wiley & Sons, Ltd, Chichester.
4. Čabrić, D., Mishra, S.M. and Brodersen, R.W. (2004) Implementation issues in spectrum sensing for cognitive radios, in The Thirty-Eighth Asilomar Conference on *Signals, Systems and Computers*, 2004.
5. Čabrić, D. and Brodersen, R.W. (2005) Physical layer design issues unique to cognitive radio systems, in IEEE 16th International Symposium on *Personal, Indoor and Mobile Radio Communications*, 2005.
6. Qiu, R., Nan Guo, N., Husheng Li, H., Wu, Z., Chakravarthy, V., Song, Y., Hu, Z., Zhang, P. and Chen, Z. (2009) A unified multi-functional dynamic spectrum access framework: tutorial, theory and multi-GHz wideband testbed. *Sensors*, **9**, 6530–6603; doi: 10.3390/s90806530.
7. Maharjan, S., Po, K. and Takada, J.-I. (2007) Energy Detector Prototype for Cognitive Radio System. IEICE Technical Report.
8. Zeng, Y., Koh, C.L. and Liang, Y. (2008) Maximum eigenvalue detection: theory and application, in IEEE International Conference on *Communications (ICC)*, May 2008.
9. Zeng, Y. and Liang, Y. (2007) Covariance based signal detections for cognitive radio, in *Proceedings of 2nd IEEE International Symposium on New Frontiers in Dynamic Spectrum Access Networks*, Chicago, IL, 17–20 April 2007, pp. 202–207.
10. Wei, L. and Tirkkonen, O. (2011) Analysis of scaled largest eigenvalue based detection for spectrum sensing, in IEEE International Conference on *Communication*, 2011.

11. Nadler, B., Penna, F. and Garello, R. (2011) Performance of eigenvalue-based signal detectors with known and unknown noise level, in IEEE International Conference on *Communication*, 2011.

12. Wax, M. and Kailath, T. (1985) Detection of signals by information theoretic criteria. *IEEE Transactions on Acoustics, Speech and Signal Processing*, **ASSP-33** (2), April, 387–392.

13. Kay, S.M. (1998) *Fundamentals of Statistical Signal Processing: Detection Theory*, vol. 2, Prentice Hall.

14. Kim, K., Akbar, I.A., Bae, K.K., Um, J.-S., Spooner, C.M. and Reed, J.H. (2007) Cyclostationary approaches to signal detection and classification in cognitive radio, in 2nd IEEE International Symposium on *New Frontiers in Dynamic Spectrum Access Networks*, 2007.

15. Po, K. and Takada, J.-I. (2007) Signal Detection Method Based on Cyclostationarity for Cognitive Radio. IEICE Technical Report.

16. Baldo, N., Asterjadhi, A. and Zorzi, M. (2009) Cooperative detection and spectrum reuse using a network coded cognitive control channel. *IEEE Network SI on Multi-Hop Cognitive Radio Networks*, **23** (4).

17. Unnikrishnan, J. and Veeravalli, V.V. (2007) Cooperative spectrum sensing and detection for cognitive radio, in *IEEE Global Telecommunications Conference*, 2007.

18. Mishra, S.M., Sahai, A. and Brodersen, R.W. (2006) Cooperative sensing among cognitive radios, in IEEE International Conference on *Communications, ICC 2006*, vol. 4, June 2006, pp. 1658–1663.

19. Visotsky, E., Kuffner, S. and Peterson, R. (2005) On collaborative detection of TV transmissions in support of dynamic spectrum sharing, in First IEEE International Symposium on *New Frontiers in Dynamic Spectrum Access Networks, DySPAN 2005*, November 2005, pp. 338–345.

20. Cao, L., Yang, L. and Zheng, H. (2010) The impact of frequency-agility on dynamic spectrum sharing, in IEEE International Symposium on *New Frontiers in Dynamic Spectrum Access Networks, DySPAN 2010*.

21. QinetiQ, Ltd. (2006) A Study of the Provision of Aggregation of Frequency to Provide Wider Bandwidth Services. Final report for Office of Communications (Ofcom), UK, QINETIQ/06/01773, August 2006.

22. Macrothink Institute (2010) *Spectrum Aggregation: Overview and Challenges, Network Protocols and Algorithms*, vol 2, no. 1; ISSN 1943-3581 2010.

23. Chen, D., Zhang, Q. and Jia, W. (2008) Aggregation aware spectrum assignment in cognitive ad-hoc networks, in 3rd International Conference on *Cognitive Radio Oriented Wireless Networks and Communications*.

24. Wang, X., Yang, C. and Zhou, J. (2008) Spectral aggregation for clustering ensemble, in 19th International Conference on *Pattern Recognition*.

25. Etkin, R. (2006) Spectrum Sharing: Fundamental Limits, Scaling Laws, and Self-Enforcing Protocols. EECS Department, University of California, Berkeley Technical Report UCB/EECS 2006-168, 11 December 2006.

26. Etkin, R., Parekh, A. and Tse, D. (2007) Spectrum sharing for unlicensed bands. *IEEE Journal on Selected Areas in Communications*, **25** (3), April, 517–528.

27. Roxin, A., Gaber, J., Wack, M. and Nait-Sidi-Moh, A (2007) Survey of wireless geolocation techniques,in IEEE Globecom Workshops.

28. Giaconi, M., Iacobucci, M.S., Semenzato, P. and Guidotti, M. (2000) An easy way to teach digital transmissions, in *Proceedings of the 11th Annual Conference on the EAEEIE*, 26–28 April 2000, Ulm, Germany.

29. Shannon, C.E. (1948) A mathematical theory of communication. *The Bell System Technical Journal*, **27**, July, 379–423; October, 623–656.

30. Lin, Z., Erkip, E. and Ghosh, M. (2005) Adaptive modulation for coded cooperative systems, in IEEE 6th Workshop on *Signal Processing Advances in Wireless Communications*, 2005.

31. Mardani, M., Harsini, J.S., Lahouti, F. and Eliasi, B. (2008) Joint adaptive modulation-coding and cooperative ARQ for wireless relay networks, in IEEE International Symposium on *Wireless Communication Systems*, 2008.

32. Liu, P., Tao, Z., Lin, Z., Erkip, E. and Panwar, S. (2006) Cooperative wireless communications: a cross layer approach. *IEEE Wireless Communications*, **13** (4), August, 84–92.

33. Hamdi, K. and Letaief, K.B. (2009) Cooperative communications for cognitive radio networks. *Proceedings of the IEEE*, **97** (5), May, 878–893.

34. Hamdi, K., Zhang, W. and Letaief, K.B. (2007) Power control in cognitive radio systems based on spectrum sensing side information, in IEEE International Conference on *Communications*, 2007.

35. Srinivasa, S. and Jafar, S.A. (2010) Soft sensing and optimal power control for cognitive radio. *IEEE Transactions on Wireless Communications*, **9** (12), December, 3638–3649.

36. Hoven, N. and A. Sahai, A. (2005) Power scaling for cognitive radio, in *Proceedings WCNC*, Maui, HI, June 2005, vol. 1, pp. 250–255.

3

Self-Organizing Network Features in the 3GPP Standard

3.1 Self-Organizing Networks

A self-organizing network (SON) is a network able to self-adjust configuration parameters to optimize network performances. Advanced network management features are introduced in order to automate the configuration of radio network parameters to improve performances, reduce faults, reduce operation and maintenance costs, and optimize the network equipment, coverage and energy consumption. The concept of SON has been introduced in release 8 of the 3GPP standard [1], in line with the vision of the Next Generation Mobile Networks (NGMN) association.

The NGMN Alliance is 'a forum to share, assess and drive aspects of mobile broadband technologies focusing on LTE & EPC (evolved packet core) and its evolution'. NGMN provides recommendations and key use cases on SON and O&M (operation and maintenance) requirements [2–4], being a guidance for 3GPP technical specifications. NGMN gives a high level description of use cases, considered from 3GPP for technical specifications emissions. Examples of use cases are: automatic optimization of radio parameters, alarming, operational support system (OSS) automation of processes and applications, and energy saving.

Self-organizing networks can be distinguished by the following groups:

- Distributed, if SON algorithms are implemented at the level of element managers or inside the network elements.
- Centralized, if SON functionalities are provided by the existing network management system or reside in an external management node. Centralized SON functionalities are implemented at the network management level.
- Hybrid, if SON algorithms are implemented both at the element management level and at the network management level.

In distributed SONs, each network node is carried by additional management functions and makes autonomous optimization decisions. The management system only sets high level SON

Reconfigurable Radio Systems: Network Architectures and Standards, First Edition. Maria Stella Iacobucci.
© 2013 John Wiley & Sons, Ltd. Published 2013 by John Wiley & Sons, Ltd.

policies and the signalling between network nodes and the management system is minimized. The times needed to collect measurements, analyse and apply optimization policies are fast, but network instabilities can occur.

In centralized SONs, the self-organizing functionalities are centralized in the management system. In this case the network nodes are relieved from a computational point of view, but additional signalling between the network management system and the network nodes is needed, and related delays in SON decisions must be taken into account. Hybrid SONs are in the middle, while part of SON algorithms are distributed in the network nodes and part are centralized in the management system.

In centralized and hybrid SONs, multiple target optimizations are possible and the operator is able to control the whole optimization process.

Figure 3.1 shows the distributed and centralized SON concepts.

Today, operators have to manage different 3GPP radio access technologies (RATs), like GSM/GPRS/EDGE, UMTS/HSPA/HSPA+ and LTE, working at different radio frequencies and with multiple layer coverage with macro, micro, pico and femto cells. The goal is to optimize the management of the whole radiomobile network, through a centralized optimization process working in a multivendor, multiradio access technology scenario. The optimization process provides the support of a centralized optimization entity, which collects measurements and performs a closed loop process in which it chooses the parameters, evaluates performances and returns the optimized parameters to the network nodes.

Figure 3.2 shows a possible architecture of a multiradio access technology network, with access network nodes administered by element managers (EMs), in general vendor dependent, connected to a centralized optimization entity, the network manager (NM), which jointly controls the distributed process of elements management. The element managers with the network manager form the operational support system (OSS). Among the access network nodes, only LTE network elements implement self-optimization algorithms, described in Section 3.3 of this chapter.

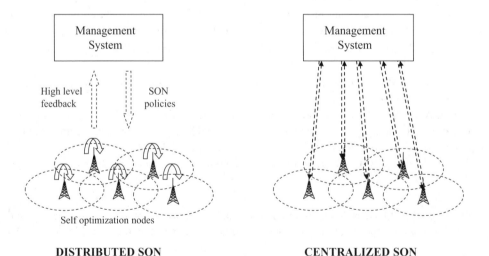

Figure 3.1   Concept of distributed and centralized SON.

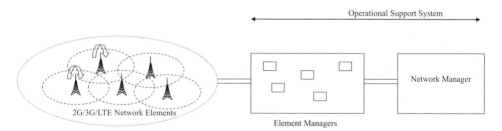

Figure 3.2 Example of management architecture for a multiradio access technology network.

3.1.1 Alarming

The task of identifying and correcting faults is increasingly difficult in a growing multi-technology access network scenario. Faults in a telecommunication network are reported to management centres in the form of alarms generated by the network nodes. Very often a fault is correctly identified through different alarms generated from different nodes. Some information carried by different alarms is redundant, others are unnecessary.

Alarms must be designed taking into account the network structure (a single node is not aware of the network architecture), they must be event triggered (sent only if needed) and alarm correlation must be taken into account for a correct identification of the fault. In general, in the network the number of alarms should be small but meaningful. This goal can be reached through the introduction of a processing function, which suppresses redundant alarms, makes alarms correlate, handles events arriving out of order, in general processes a large number of alarms and reduces them in a few meaningful pieces of information sent to the management center (MC). The processing function could be external or internal to the MC. The MC generates an automatic resolution of the fault. This process is shown in Figure 3.3. The management centre is at the NM level and the EM level. The operator maintains control of fault resolution by establishing at the NM level the policies for alarm reporting and fault resolution.

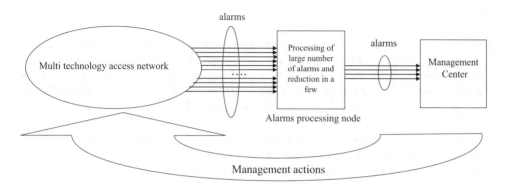

Figure 3.3 Example of alarm processing and fault resolution.

3.1.2 Operational Support System Automation

Operational support system (OSS) automation helps to reduce operating costs, increase operational efficiency and enhance quality of services. In a multitechnology access network scenario a common OSS could help to manage all the network elements, of both access and core networks. In this scenario, a centralized management application implements the policies and manages activities like a network element health check, software downloads, software upgrades and corrective actions. Network changes of hardware and software should also automatically change the characteristics of alarms and their processing.

A very interesting evolution of OSS is cloud OSS. The model of cloud computing extends the service-oriented approach to the 'cloud' of the network. Cloud computing is emerging as a model that radically changes the provisioning of hardware and software resources [5]. The architecture is *multitenant*, that is the resources are shared among different users. A definition of cloud has been given by the National Institute of Standards and Technology (NIST) in September 2011: 'Cloud computing is a model for enabling ubiquitous, convenient, on-demand network access to a shared pool of configurable computing resources (e.g., networks, servers, storage, applications, and services) that can be rapidly provisioned and released with minimal management effort or service provider interaction' [6].

Cloud computing includes all the disciplines, technologies and business models using IT (information technology) resources stored in servers connected to the Internet, according to an on demand, pay per use model. IT resources can be applications, software services, computing platforms, storage systems, etc.

Cloud computing is therefore a new style of conceiving the supply of IT services through the convergence of three key elements:

- Utility computing: the use of computational resources is considered as an always available, always measurable commodity, adapting to the user needs with a pay per use model.
- Computational resources virtualization: virtualization abstracts the underlying physical infrastructure through the creation of logical resources. It is an enabling characteristic of cloud computing because it reduces times and costs of provisioning, and moves part of the costs of computer equipment from CAPEX to OPEX.
- Software as a service: is the new way of considering software applications as services, reusable and interoperable.

Cloud computing can also be identified as:

- *Private cloud:* enables the delivery of services to a selected group of consumers. A model of private cloud is *community cloud.* In this case services are offered to users having common interests and regulated by appropriate service level agreements (SLAs) or security policies. Private cloud may be owned, managed and operated by the organization that uses the private cloud services, a third party or some combination of them, and it may exist on or off premises [6].
- *Public cloud:* is maintained from a cloud service provider and enables the delivery of services to all customers though the public Internet.
- *Hybrid cloud:* includes the characteristics of private and public cloud.

Figure 3.4 shows an example of private, public and hybrid cloud.

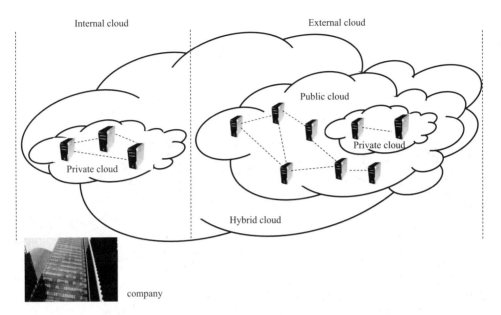

Figure 3.4 Example of private, public and hybrid cloud.

If the network management application of an operational support system is put into the cloud, this is cloud OSS. Operation support system cloud services can provide the right combination of technology and flexibility, for example with dynamic OSS resource usage with different network loads.

An OSS must be a private cloud, because its users are the element managers of the network nodes. Big telcos will perhaps internally manage OSS clouds, while small or medium telcos will have the opportunity to use on demand cloud OSS services, fully web enabled and secure, through reduced operational expenses and in a pay per use model. A cloud OSS requires real time processing. This model is shown in Figure 3.5.

3.1.3 Energy Saving

For a telecommunication company a large amount of operational costs is related to energy consumption. In the past, some low traffic nodes were switched off during the night for saving cost reasons, even if they would have lost some traffic.

An automation of energy saving is fundamental for costs optimization without traffic losses. With energy saving automation, a network node will be able adaptively to switch off or put in standby mode the unused resources, in order to consume the minimum amount of energy required to serve the actual users. These resources can be a certain number of subcarriers, processing modules or transmission links, possibly managed with a fine granularity in order to match perfectly the working resources to the amount of traffic.

Together with activation or deactivation of network resources, a new parameterization of the involved node and of its neighbours must be automatically performed. For example, if

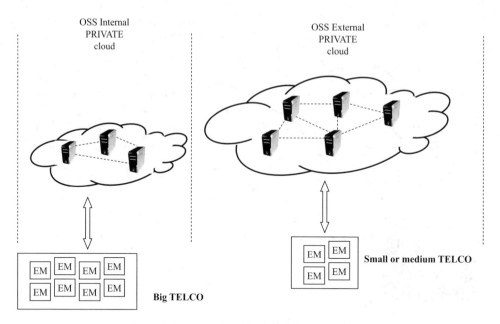

Figure 3.5   Example of OSS private cloud services.

some subcarriers are switched off because of low traffic, logical channel mapping must be updated. If a cell or a node is completely switched off, the neighbours will have to rearrange the radio parameters like output power, electrical tilt and handover thresholds, to guarantee the service the requested quality. The decision of switching off/on some resources is based on traffic thresholds.

In a multiradio access technology scenario, the optimization of radio parameters involves cells of different technologies. This is why the energy saving process should be controlled by a centralized optimization centre.

A possible architecture and flow diagram for radio energy saving policies is shown in Figure 3.6. In the example there are different access nodes, like LTE eNBs, UMTS NBs, GSM BTSs. All the nodes periodically send traffic reports to a management centre (MC), which decides if some radio resources can be switched off. After the decision, the management centre informs each node about the resources that can be switched off and updates the radio parameters.

The first step of the flow diagram, which is the evaluation of traffic reports and the identification of low traffic periods and that in the figure is shown to be in the MC, could also be done in each access node; only traffic reports that are under/above the threshold could be sent to the MC. In this case the amount of signalling to the MC other than its processing resources would be much lower. In this partially distributed approach the management centre sends traffic threshold adjustments to the nodes if required.

There are some attention points to be taken into account, like maintaining QoS requirements when resources are switched off and maintaining the stability of the network when parameters are changed. The management centre is at the network manager (NM) level and at the element

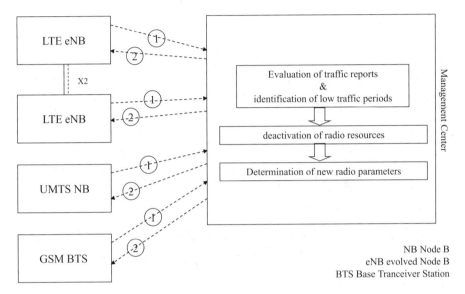

Figure 3.6   Example of architecture for energy saving through radio resources deactivation.

manager (EM) level. The operator can maintain control of the energy saving process at the network manager level by establishing the policies for reporting, energy saving actions and radio network new parameterization.

3.2 LTE Overview

The LTE network architecture and the radio interface have been described in Chapter 1. In this section the LTE network architecture and radio features are summarized.The LTE network architecture consists of an access network, the evolved UTRAN (E-UTRAN), connected to a core network, the evolved packet core (EPC). The LTE network architecture is shown in Figure 3.7.

The E-UTRAN presents a flat architecture made of interconnected evolved node Bs (eNBs). The main functionalities performed by an eNB are radio resource management (RRM), mobility management (MM), IP header compression and encryption of user data, selection of an MME at a UE attachment, routing of user data to a gateway, scheduling and transmission of control messages (paging, broadcast), measurement and measurement reporting configuration for mobility and scheduling.

The radio interface is based on orthogonal frequency division multiple access (OFDMA) in downlink and single carrier FDMA (SC-FDMA) in uplink. Duplexing can be both time division duplex (TDD) and frequency division duplex (FDD); the bandwidth is variable form

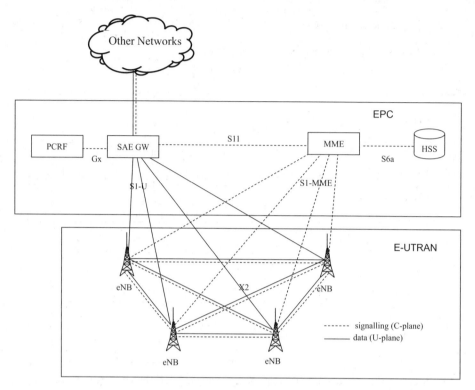

Figure 3.7   LTE network architecture.

1.4 MHz up to 20 MHz. The minimum allocable radio resource is the resource block (RB), which is a group of 12 subcarriers each 15 kHz wide, in a time of one millisecond. Multiple input multiple output (MIMO) antennas are supported in downlink, in the configuration 2×2 and 4×4. Adaptive modulation and coding is implemented, with bit rates depending on bandwidth, MIMO configuration, modulation and code rate. Table 3.1 shows how downlink physical bit rates depend on transmission parameters; Table 3.2 shows how uplink physical bit rates depend on transmission parameters.

The modulation and coding scheme (MCS) is chosen on the basis of the radio channel, which is estimated in uplink from the eNB and in downlink from the mobile station (MS). The assignment of downlink radio resources as a function of the radio conditions is performed through the *channel feedback reporting procedure*. During the eNB transmission, the MS measures the downlink channel and sends, as a feedback, the channel quality indicator (CQI), corresponding to the modulation and coding scheme (MCS) and transport block size (TBS) for which the estimated received downlink transport block error rate (BLER) shall not exceed 10%. TBS is the amount of data carried in a TTI.

The rank indicator (RI) and precoding matrix indicator (PMI) are feed-backed parameters for MIMO. RI indicates the rank of the estimated channel matrix **H** and equals the number of usable spatial streams. It is related to the whole system bandwidth. PMI indicates the preferred precoding matrix in a codebook if a closed loop MIMO configuration is adopted.

Table 3.1 LTE downlink physical bit rates

Modulation	Code rate	MIMO	Bandwidth (MHz)						
			1.4	3	5	10	15	20
QPSK	1/2	Single stream	0.8 Mbps	2.2 Mbps	3.7 Mbps	7.4 Mbps	11.2 Mbps	14.9 Mbps
16-QAM	1/2	Single stream	1.5 Mbps	4.4 Mbps	7.4 Mbps	14.9 Mbps	22.4 Mbps	29.9 Mbps
16-QAM	3/4	Single stream	2.3 Mbps	6.6 Mbps	11.1 Mbps	22.3 Mbps	33.6 Mbps	44.8 Mbps
64-QAM	3/4	Single stream	3.5 Mbps	9.9 Mbps	16.6 Mbps	33.5 Mbps	50.4 Mbps	67.2 Mbps
64-QAM	1	Single stream	4.6 Mbps	13.2 Mbps	22.2 Mbps	44.7 Mbps	67.2 Mbps	89.7 Mbps
64-QAM	3/4	2×2 (spatial multiplexing)	6.6 Mbps	18.9 Mbps	31.9 Mbps	64.3 Mbps	96.7 Mbps	129.1 Mbps
64-QAM	1	2×2 SM	8.8 Mbps	25.3 Mbps	42.5 Mbps	85.7 Mbps	128.9 Mbps	172.1 Mbps
64-QAM	1	4×4 SM	16.6 Mbps	47.7 Mbps	80.3 Mbps	161.9 Mbps	243.5 Mbps	325.1 Mbps

Table 3.2 LTE uplink physical bit rates

Modulation	Code rate	MIMO	Bandwidth (MHz)					
			1.4	3	5	10	15	20
QPSK	1/2	Single stream	0.9 Mb/s	2.2 Mb/s	3.6 Mb/s	7.2 Mb/s	10.8 Mb/s	14.4 Mb/s
16-QAM	1/2	Single stream	1.7 Mb/s	4.3 Mb/s	7.2 Mb/s	14.4 Mb/s	21.6 Mb/s	28.8 Mb/s
16-QAM	3/4	Single stream	2.6 Mb/s	6.5 Mb/s	10.8 Mb/s	21.6 Mb/s	32.4 Mb/s	43.2 Mb/s
16-QAM	1	Single stream	3.5 Mb/s	8.6 Mb/s	14.4 Mb/s	28.8 Mb/s	43.2 Mb/s	57.6 Mb/s
64-QAM	3/4	Single stream	3.9 Mb/s	9.7 Mb/s	16.2 Mb/s	32.4 Mb/s	48.6 Mb/s	64.8 Mb/s
64-QAM	1	Single stream	5.2 Mb/s	13 Mb/s	21.6 Mb/s	43.2 Mb/s	64.8 Mb/s	86.4 Mb/s

In the evolved packet core (EPC), signalling and data are separated and managed by different nodes. The mobility management entity (MME) is connected to the eNBs through the S1-MME interface, which carries control plane messages. The system architecture evolution gateway (SAE GW) is connected to the eNBs through the S1-U interface, which carries user plane messages. The home subscriber server (HSS) is the repository of all permanent user data, like the subscriber profile and the permanent key for authentication, ciphering and integrity protection. It also stores the location of the user at the level of the visited network control node, such as MME. The MME manages mobility, MS identities and security parameters.

The SAE gateway can be separated into two gateways: the serving gateway (SGW) and the packet gateway (PGW). The primary task of the SGW is IP routing and forwarding of IP packets. It is the user plane anchor for inter-eNB handover (HO) and when the user moves among 3GPP access technologies. The PGW is connected to the SGW and to the external networks. It is responsible for the QoS and is the user plane anchor when the MS moves among 3GPP and non-3GPP radio access technologies. One MS can have simultaneous accesses to more than one PGW (e.g., to have access to different packet data networks). The SAE gateway also performs packet filtering and IP address allocation to mobile stations.

The last element included in the EPC is the policy and charging resource function (PCRF), connected with the Gx interface (signalling only) to the PGW. It implements policy and charging rules and elaborates policy and charging control requests.

The LTE core network is supposed to be connected through standard interfaces to the 2G/3G networks, as shown in Figure 3.8. Thanks to the standard interfaces, circuit switched fallback (CSFB) and inter-RAT (radio access technology) handovers can be performed. Circuit switched fallback (CSFB) was introduced in release 8 and allows a mobile station (MS)

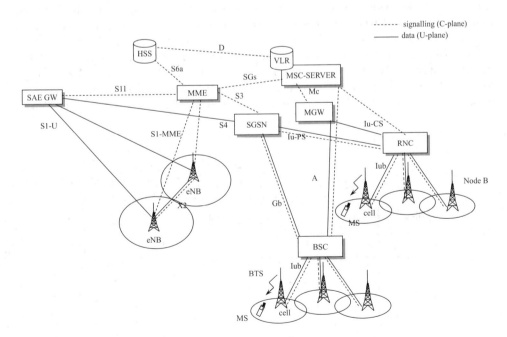

Figure 3.8   2G/3G/4G network architecture with interfaces.

registered in the LTE domain to use circuit switched services in the GSM/GPRS/EDGE or UMTS/HSPA/HSPA+ domain.

The evolution of LTE, which is LTE advanced, has been standardized starting from release 10 of the 3GPP standard and introduces new radio features to improve the spectral efficiency in uplink and downlink and to increase the cell edge throughput. The principal features introduced in LTE advanced are:

- Support of wider bandwidth: the aggregation of two or more component carriers, each with a bandwidth up to 20 MHz, is considered for LTE advanced in order to support transmission bandwidths larger than 20 MHz, up to 100 MHz.
- Extended MIMO configurations: up to an 8×8 MIMO configuration is considered in downlink; up to a 4×4 MIMO configuration is considered in uplink.
- Coordinated multiple point (CoMP) transmission and reception: this feature improves high data rate coverage and cell-edge throughput. Downlink coordinated multipoint transmission implies dynamic coordination among multiple geographically separated transmission points.
- Relaying: used to improve the coverage of high data rates, cell-edge throughput. Provides coverage in new areas.

Carrier aggregation and extended MIMO configuration increase spectral efficiency, while CoMP and relaying enhance cell-edge throughput and coverage.

3.3   LTE Home eNB

A home eNB (HeNB) is an LTE femto cell. A femto cell is a low power home base station, connected to the broadband access network. The femto cells are used for indoor coverage, to provide voice and data services. The network architecture is represented in Figure 3.9.

The eNBs are connected to an eNB gateway, which is seen from the eNBs as a mobility management entity. The security gateway allows a security tunnel establishment for the backhaul link. The home eNB (HeNB) is connected to a broadband access device, which guarantees the broadband connection with the access network. The broadband access device can also be

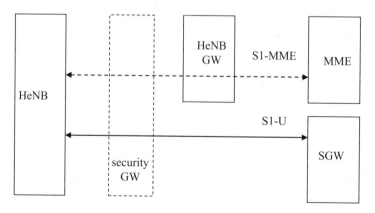

Figure 3.9   Home eNB network architecture.

integrated with the HeNB. Examples of broadband access devices are optical line terminations (OLTs), VDSL modems, etc. HeNB functionalities and interfaces are the same as eNBs, with some added functions.

Access policies for HeNBs are based on the eNB cell identifier and, according to release 8 of the 3GPP standard, can be:

- Closed: mainly for residential users. The HeNB is defined as a closed subscriber group (CSG) cell and the access control is located in the gateway.
- Open: the access to the HeNB is allowed to all users.

In release 9 of the 3GPP standard a hybrid access policy has been introduced. In this case the users belonging to the CSG can be prioritized and differently charged from the unsubscribed users.

Because HeNBs are in general located at the customer's premises, their configuration and maintenance must be handled with a remote connection. Management functionalities and interfaces are standardized to support multivendor environments. Figure 3.10 shows the HeNB management architecture, composed of an HeNB management system connected through a type 1 interface to many HeNBs.

The HeNB management system (MS) handles different functions, such as:

- Discovery and assignment of the HeNB management system,
- Identity and location verification of HeNBs,

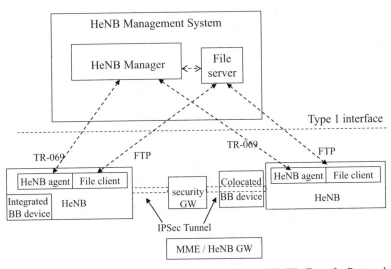

FTP File Transfer Protocol

Figure 3.10   HeNB management architecture. (Reproduced with permission. Copyright 2011. 3GPP™ TSs and TRs are the property of ARIB, ATIS, CCSA, ETSI, TTA and TTC who jointly own the copyright in them. They are subject to further modifications and are therefore provided to you "as is" for information purposes only. Further use is strictly prohibited.)

- Software downloading from the MS to the HeNBs,
- Key performance indicator (KPI) provision from the HeNB for performance management (PM), and
- HeNB configuration management (CM), fault management (FM), performance management (PM).

Procedure flows for type 1 interface have been introduced in release 9 of the 3GPP standard and enhanced in releases 10 and 11 [7].

When the HeNB is switched on, it has to establish a connection with the home eNBs management system (HeMS) and with the other network nodes, like the mobility management entity (MME). The Internet protocol is used for communication among network nodes. The HeNB must be able to acquire the IP address of the HeMS and, through it, the IP address of the MME. At this point the S1 interface is set up. The security gateway provides a secure connection between HeNB and HeMS, and between HeNB and MME.

A file server is implemented in the HeMS and the file client is in the HeNBs. The client–server file architecture includes, as an optional feature, the HeNB configuration using a file download procedure. The HeNB discovery and registration procedure is shown in Figure 3.11 [7].

The application layer protocol for remote management of HeNB has been defined from the broadband forum in Technical Report 069 (TR-069) [8]. It handles functions such as:

- Autoconfiguration and dynamic services activation: initial and remote configuration of the HeNBs.
- HeNBs firmware management: version and update management.
- HeNBs status and performances monitoring: analysis of log files, diagnostic services, connection control, and
- Diagnostics, alarming.

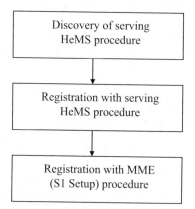

Figure 3.11   Discovery and registration procedure. (Reproduced with permission. Copyright 2011. 3GPP™ TSs and TRs are the property of ARIB, ATIS, CCSA, ETSI, TTA and TTC who jointly own the copyright in them. They are subject to further modifications and are therefore provided to you "as is" for information purposes only. Further use is strictly prohibited.)

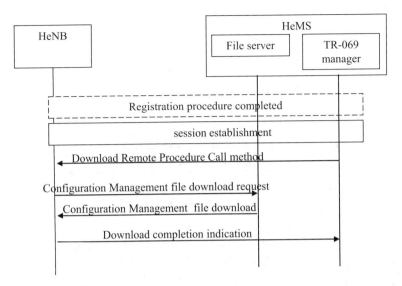

Figure 3.12 HeNB configuration using the file download procedure (optional). (Reproduced with permission. Copyright 2011. 3GPP™ TSs and TRs are the property of ARIB, ATIS, CCSA, ETSI, TTA and TTC who jointly own the copyright in them. They are subject to further modifications and are therefore provided to you "as is" for information purposes only. Further use is strictly prohibited.)

The HeNB manager that resides in the HeNB management system is a TR-069 manager. The HeNB configuration using the file download procedure (optional) is shown in Figure 3.12 [7].

After completion of the registration procedure, a TR-069 session is established between the eNB and the TR-069 manager. The TR-069 manager sends the remote procedure call (RPC) method that must be used, including the file type, URL of the source file location, user name and password for the connection to the file server, as well as the file size [8]. After the request, the configuration management (CM) file is downloaded and the procedure is completed with a notification.

The home eNB configuration can also be performed by using the SetParameterValues method (mandatory), invoked from the eMS to configure the parameters in the HeNB. The new parameters sent from the manager to the HeNB are applied and a response is sent back. The described procedure is shown in Figure 3.13 and can be run or can be used when the registration procedure is completed to configure the HeNB, or after a procedure of alarm reporting or performance management (PM) reporting. In the latter case the parameter configuration procedure is used to solve the fault that generated the alarm or to optimize the performances after PM reporting. The configuration procedure can also be run for periodic parameter updating.

3.4 LTE and Self-Organizing Networks

LTE and EPC have been introduced in release 8 of the 3GPP standard. Self-organizing network (SON) features have been included from the first release of the standard, and have been updated and extended in the subsequent releases. The reason why self-organizing networks were taken

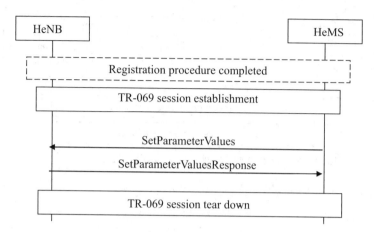

HeNB parameter configuration procedure using
SetParameterValues method

Figure 3.13 HeNB parameter configuration and updating procedure using the SetParameterValues method (mandatory). (Reproduced with permission. Copyright 2011. 3GPP™ TSs and TRs are the property of ARIB, ATIS, CCSA, ETSI, TTA and TTC who jointly own the copyright in them. They are subject to further modifications and are therefore provided to you "as is" for information purposes only. Further use is strictly prohibited.)

into account from the beginning is that the rapid evolution of radiomobile networks leads to a multiple radio access technology network with many and complex parameters to plan and manage. SONs meet the needs of operators that have to configure and optimize the parameterization of all the network nodes with minimum human intervention. Figure 3.14 shows the principal SON functionalities provided starting from release 8 up to release 11 of the 3GPP standard [9].

Concepts and requirements of self-organizing networks have been defined in release 8 [10] and updated in subsequent releases [11]. Requirements provide a scenario where the network can be shared among multiple operators. SON functions will be at the beginning operator controlled (open loop), and only when the operator trusts the good functioning of self-organizing features will it gradually switch to closed loop, which means that the network will autonomously operate under the control of SON functionalities. Requirements also enable SON operations and provide SON capabilities inside the operational support system (OSS), and define interfaces for the following functionalities:

- Self-establishment of a new eNB in the network
- Automation of neighbour relation lists in E-UTRAN, UTRAN and between different 3GPP radio access technologies
- Self-configuration and self-healing of eNBs
- Automated coverage and capacity optimization
- Optimization of parameters due to troubleshooting
- Continuous optimization due to dynamic changes in the network

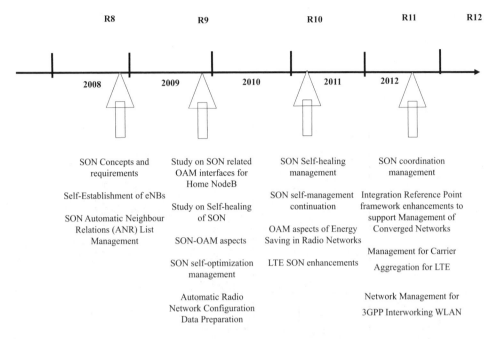

Figure 3.14 LTE SON principal functionalities in 3GPP releases.

- Automated handover optimization
- Optimization of QoS related radio parameters

3.4.1 Self-Establishment of a New eNB

Concepts and requirements for self-establishment of a new eNB have been identified in release 8 [12], and have been updated and extended in the subsequent releases to the concept and requirements for self-configuration of network elements [13]. The scenario is the LTE access network of numerous eNBs interconnected through the X2 interface and connected with the EPC nodes through the S1 interface. The eNBs can be from different vendors and could be shared among different operators. The reference model for self-configuration is made of numerous functional blocks, and is shown in Figure 3.15.

The self-configuration monitoring and management function (SC-MMF), which controls the self-configuration process, is split into two parts. The first part is implemented at the element manager level and is called the self-configuration monitoring and management function element manager (SC_MMF_EM); it is able to get information and give feedback to all other functional blocks. The second part is implemented at the network manager level and is called the self-configuration monitoring and management function network manager (SC_MMF_NM); it is connected to the SC_MMF_EM with the standard interface Itf-N, and allows the operator to control the execution of the self-configuration process [13]. At the network manager level all the functions requiring coordination among several eNBs must be implemented.

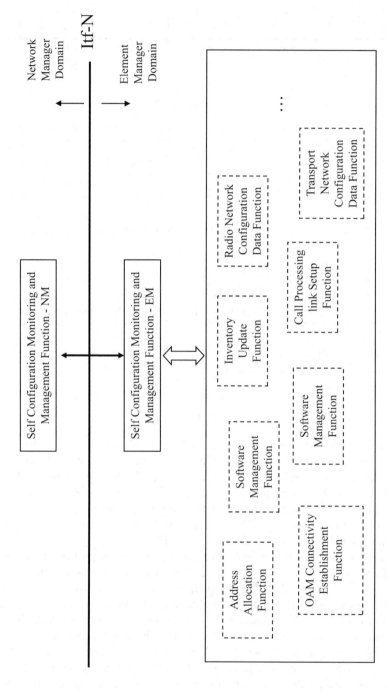

Figure 3.15 Self-configuration reference model.

When the eNB is switched on, it authenticates the network and vice versa. In this case an external planning tool has determined the eNB location and the hardware for installation including antennas. An initial group of parameters is set up, like the eNB type, cell characteristics (sectors), etc.

After installation, the eNB will run a self-test, with eventual failure reporting to the element manager. If the self-test results are positive, then the process of eNB self-configuration starts with the detection of the operational support system (OSS), from which it eventually downloads the most recent software. After that, the network parameters to establish a logical connection to the network and the OSSs are set up. At this time the eNB and the connected nodes have an IP address and have set QoS parameters of the transport interface.

After that it determines the initial radio configuration. It sets the output power, the electrical tilt for coverage, configures physical resources and maps logical channels, generates the cell identifier (Cell-Id), downloads the neighbour list from an external database, sets up radio parameters for congestion and admission control, handover thresholds, etc. At the end of this process various tests are performed in order to assure the proper functioning of eNB [14].

After the self-configuration, a continuous self-optimization process is executed. The self-optimization process is described in Section 3.4.3 and involves the following features:

- Coverage optimization: tilt and output power adjustments.
- Neighbour list updating: the eNB identifies new neighbours through power measurements and erases the weaker neighbours. The X2 interface with the new neighbours is set up and the same interface with the old neighbours is slowed down. The automatic neighbour relation management is described in Section 3.4.2.
- Handover optimization: handover related measurements, like call drops, and handover failures are used to adjust handover thresholds and hysteresis in order to minimize the failures.
- Interference control: the radio resource management (RRM) assigns resource blocks to minimize the global system interference.
- QoS parameter optimization: QoS related measurements, like average cell throughput, average throughput per user, average delay per user and cell load, are used to adjust QoS related parameters.
- Cell outage compensation: the network reconfigures radio parameters in order to avoid loss of service. For example, the neighbour cell parameters, like antenna tilt, power and channel configuration, are modified in order to compensate the loss of coverage and service. The neighbour list shall be updated. X2 interfaces shall be reconfigured.

An example of a flow chart of eNB self-establishment is presented in Figure 3.16.

Self-establishment and self-optimization shall be supported in a multiple vendor environment. The procedures and OAM interfaces need to be standardized in order to allow complete interoperability among different vendors and to avoid side effects like network instability.

3.4.2   Automatic Neighbour Relation Management

Telecommunication operators actually need to jointly manage multi-RATs at different frequencies, with pico, micro and macro coverage. In order to enable inter-RAT handover (HO), neighbour relations have to be configured in each access network node (BTSs, NBs,

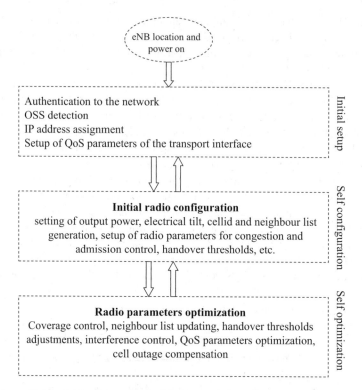

Figure 3.16   Example of a flow chart of eNB self-establishment.

eNBs). The installation of a new access network node implies the manual configuration of a new adjacency in the neighbour nodes; the removal of an access network node involves the manual removal of the neighbourhood in the adjacent nodes. Automatic neighbour relation (ANR) functionality automates the management of adjacencies in the LTE access nodes, the evolved NBs. Figure 3.17 shows the ANR function implemented in the eNB [15]. It configures the neighbour cell relations (NRs) when a new eNB is set up, and automatically optimizes the neighbourhood list.

In LTE, the ANR function resides in the eNB and manages the neighbour relation table (NRT). One neighbour relation table exists for each cell in the eNB. A neighbour cell relation (NR) from a source cell to a target cell exists if the eNB knows the following information:

- Neighbour cell is in LTE: target cell identity (TCI), which includes the E-UTRAN cell global identity (ECGI = PLMN-Id + PCI) and the physical cell identity (PCI) of the target cell,
- Neighbour cell is in UTRAN: neighbour cell identity (NCI), which is a global cell identity (GCI) including the public land mobile network (PLMN) identifier (PLMN-Id), cell identity (CI), radio network control identifier (RNC-Id), and
- Neighbour cell is in GSM/GPRS/EDGE: neighbour cell identity (NCI), which is a cell global identity (CGI) including PLMN-Id, location area code (LAC), cell identity (CI) and base station identity code (BSIC).

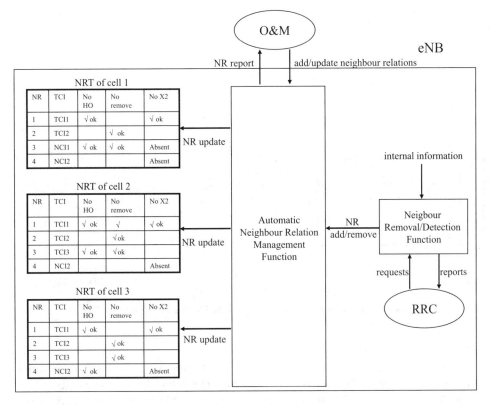

Figure 3.17   ANR function implemented in the eNB.

Other than TCI and NCI, other fields are in the neighbour relation table, controlled by the OSS system, such as:

- No X2: this field exists only for LTE neighbour cells and indicates the absence of the X2 interface between the source cell and target cell.
- No remove: the target cell cannot be removed from the NRT.
- No HO: the neighbour relation cannot be used for handover purposes.

The neighbour relation table is updated from the ANR management function, connected with the OSS system. The neighbour detection/removal function has the task of adding new neighbours and removing old ones; the information about target cells to be added or cancelled came from the radio resource control (RRC) function [16].

Therefore, the automatic neighbour relation manages intra-LTE/intrafrequency ANR, intra-LTE/interfrequency ANR and inter-RAT ANR. Both intra-LTE and inter-RAT automatic neighbour relations are controlled from the serving cell eNB, equipped with ANR functions.

In intrafrequency/intra-LTE ANR, the procedure works as follows:

1. The mobile station (MS) sends a measurement report containing the physical cell identity (PCI) of the target cell.

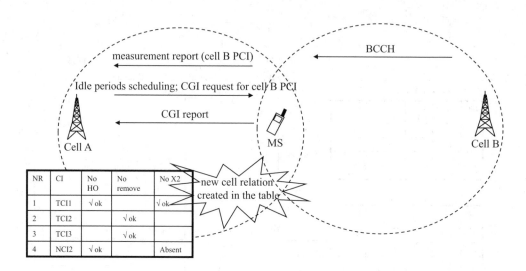

NR	CI	No HO	No remove	No X2
1	TCI1	√ ok		√ ok
2	TCI2		√ ok	
3	TCI3		√ ok	
4	NCI2	√ ok		Absent

PCI Physical Cell Identifier
CGI Cell Global Identy

Figure 3.18   Example of the intrafrequency/intra-LTE ANR procedure.

2. If the PCI is not in the NRT, the eNB schedules idle periods to allow the MS to perform new measurements with the goal to discover the target cell global identity (CGI).
3. The MS reports the ECGI, other than the tracking area code and PLMN-Id, to the serving cell, which updates its NRT.

An example of the intrafrequency/intra-LTE ANR procedure is represented in Figure 3.18.
   Each eNB contains an interfrequency search list [17], which enables inter-RAT and inter-frequency ANR. For interfrequency/inter-RAT ANR, the procedure works as follows:

1. The eNB schedules idle periods and instructs the MS to perform neighbour cell measurements in the target RAT/frequency.
2. The MS reports to the source cell the physical cell identity (PCI) of the detected cells.
3. The eNB schedules idle periods to allow the MS to perform new measurements with the goal to discover the CGI of the detected cells.
4. The MS reports the cell global identity (CGI) with other parameters like LAC (location area code) or RAC (routing area code) to the serving cell, which updates its NRT.

The PCI is not a unique identifier in radiomobile networks. This is why it is necessary to feed back other parameters to obtain a unique cell identity. Such parameters are:

• For LTE neighbour cells: E-UTRAN cell global identity (ECGI), other than the tracking area code and PLMN-Id,

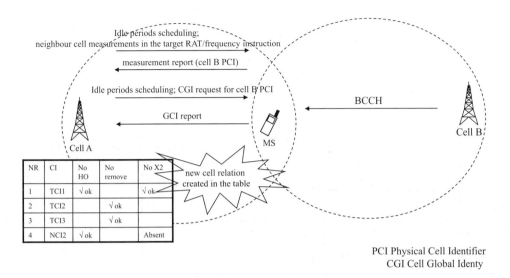

NR	CI	No HO	No remove	No X2
1	TCl1	√ ok		√ ok
2	TCl2		√ ok	
3	TCl3		√ ok	
4	NCl2	√ ok		Absent

PCI Physical Cell Identifier
CGI Cell Global Identy

Figure 3.19   Example of interfrequency/inter-RAT ANR procedure.

- For UTRAN neighbour cells: cell global identity (CGI), location area code (LAC) and routing area code (RAC), PLMN-Id, and
- For GSM/GPRS/EDGE neighbour cells: cell global identity (CGI), routing area code (RAC), PLMN-Id.

An example of the interfrequency/inter-RAT ANR procedure is shown in Figure 3.19.

3.4.3   eNB Self-Optimization

The self-optimization process regards the automatic optimization of radio parameters of the eNB in order to meet traffic requirements and optimize the access network performances. The process of self-optimization provides a self-optimization functionality that receives key performance indicators (KPIs), alarms, etc., and monitors the input data according to operator's objectives and targets. If targets are not met, the optimization algorithm is run to determine the corrective actions to be taken.

The operator may decide when to execute the proposed actions. If the corrective actions are deployed, the algorithm evaluates the obtained results and, if the targets are met, the algorithm ends and input data monitoring continues; otherwise the system could be reversed to the previous status (fallback) and after that input data are monitored. The operator may confirm the execution of fallback [18].

The described process is shown in the flow chart of Figure 3.20. The dotted lines indicate steps for operator controlled functions (open loop).

The reference model for self-optimization is made from numerous functional blocks and is shown in Figure 3.21.

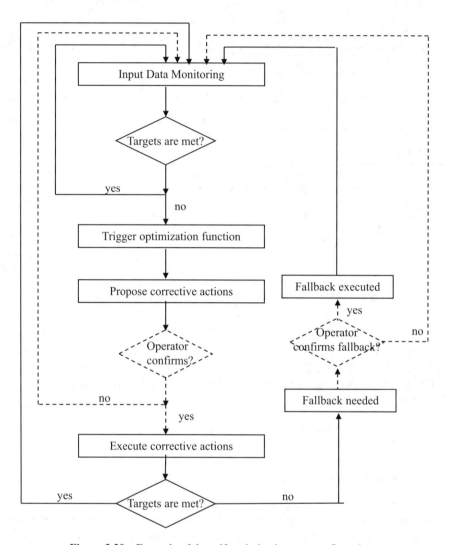

Figure 3.20   Example of the self-optimization process flow chart.

The self-optimization monitoring and management function (SO-MMF), which controls the self-optimization process, in split into two parts.

The first part is implemented at the network manager level, and is called the self-configuration monitoring and management function network manager (SO_MMF_NM). It operates above the Itf-N interface, monitors the self-configuration process and allows the operator to control the whole process so that it meets the given targets.

The second part is implemented at the element manager level, and is called the self-configuration monitoring and management function element manager (SO_MMF_EM). It operates below the Itf-N interface and identifies, resolves and/or reports conflicts, according to the policy directions received by the SO_MMF_NM.

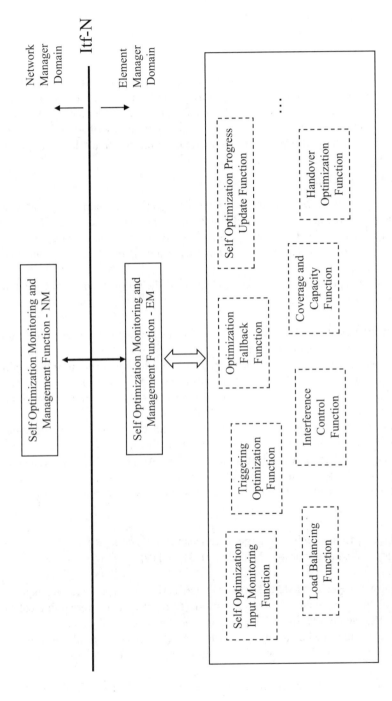

Figure 3.21 Self-optimization reference model.

Examples of policy directions are:

- SON function prioritizations in the case of conflicts,
- Inhibition of changes of one or more parameters for a certain period, and
- Selection of preferred ranges of certain values.

Each self-optimization function has several performance indicators, used to evaluate the system behaviour during the monitoring process, before and after a corrective action. Among the self-optimization functions, the use cases for load balancing, interference control, coverage and capacity optimization and handover optimization will be described [18]. All the above mentioned functions are performed with minimal human intervention (closed loop).

3.4.3.1 Load Balancing

The load balancing optimization function has the objective of distributing traffic load in order to satisfy QoS requirements and maximize the access network capacity. To activate the load balancing functionality, there must be a partial or total overlapping coverage among the involved cells. Moreover, load balancing can be actuated among intrafrequency, interfrequency or inter-RAT cells.

The load balancing optimization function is based on the following performance indicators related to traffic load:

- RRC connection establishments failure rate: indicates the rate of failures in the attempt to make an RRC connection, due to cell load.
- E-UTRAN radio access bearer (E-RAB) establishment failure: indicates the rate of failures in the attempt to establish E-RAB, due to cell load. The E-RAB, also called the evolved packet system (EPS) bearer, is a logical connection established when a mobile station connects to the core network.
- RRC connection release rate: indicates the rate of connections that suffer a degradation in the bit rate to a value lower that the target, due to traffic load.
- E-RAB abnormal release rate: indicates the rate of E-RABs that suffer a degradation in the bit rate to a value lower that the target, due to traffic load.
- Rate of failures related to handover: indicates the rate of handover failures, due to traffic load.

Operators should assign weights for target performance indicators [18].

The scenarios considered for load balancing (LB) are [19]:

- Overlapping coverage: load balancing is considered between two perfectly overlapping cells.
- Hierarchical coverage: load balancing is considered between two overlapping cells of different size. LB can be performed if the user is inside the overlapping area.
- Neighbour coverage: load balancing is considered between two partially overlapping cells of different size. LB can be performed if the user is inside the overlapping area.

Figure 3.22 shows the scenarios of (a) overlapping coverage, (b) hierarchical coverage and (c) neighbour coverage.

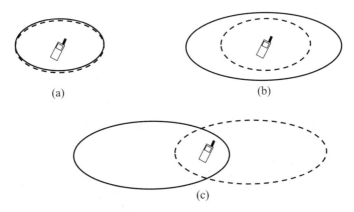

Figure 3.22 Scenarios of (a) overlapping coverage, (b) hierarchical coverage and (c) neighbour coverage.

3.4.3.2 Handover Optimization

The handover optimization function has the objective of minimizing handover failures, minimizing unnecessary handovers and increasing the load balancing capability of the network [18]. The increase of capability due to load balancing was discussed in Section 3.4.3.1. The other two points, the minimization of handover failures and the minimization of unnecessary handovers, are discussed in the present section.

To activate the handover optimization functionality, the neighbour relation tables (NRTs) must be kept up to date through the automatic neighbour relation (ANR) function discussed in Section 3.4.2. Handover optimization concerns intrafrequency, interfrequency and inter-RAT handover. The most important performance indicator for the handover optimization function is the handover related radio link failure rate. This parameter is very important because it not only affects network performances but also the user experience.

There are three main reasons for handover (HO) radio link failures:

- Too late HO triggering: it occurs if the HO is triggered when the signal strength of the source cell is too low, then causing radio link failure. The network is aware of this type of failure because the mobile station re-establishes a connection in a new cell.
- Too early HO triggering: it occurs if the HO is triggered when the signal strength of the target cell is too low, then causing radio link failure. The network is aware of this type of failure because the mobile station re-establishes a connection in the source cell.
- HO to a wrong cell. The network is aware of this type of failure because a radio link failure occurs in a cell different from the source, and the mobile station re-establishes a connection in the target cell, different from the source and from the cell where the radio link failure occurred.

Figure 3.23 shows the three cases of (a) too late HO triggering, (b) too early HO triggering and (c) HO to a wrong cell.

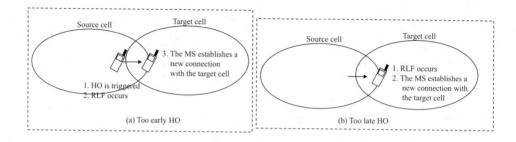

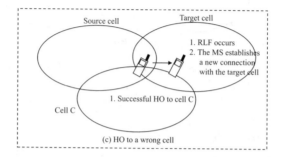

Figure 3.23   Examples of handover failures.

LTE specifies hard HO mechanisms based on reference signal received power (RSRP) measurements. The signal strength measurements are signal levels averaged over a certain amount of time.

The parameters involved in a radio link failure (RLF) that must be optimized are:

- Handover hysteresis margin (HOM): hysteresis is used instead of a threshold in the handover evaluation criteria to avoid ping-pong effects.
- Time to trigger (TTT): identifies the time during which specific criteria for the event need to be met in order to trigger a measurement report [20].
- Cell individual offset (CIO): is an added offset for handover determination. It is used to increase or decrease the handover areas.
- Cell reselection parameters: are the parameters taken into account for the cell reselection procedure. Some of them are: cell selection RX level value, cell selection quality value, etc.

Figure 3.24 shows an example of the HO evaluation algorithm in the case of (a) successful handover and (b) radio link failure. Hysteresis, TTT and RLF threshold parameters are shown in the figure. It is shown how a larger TTT determines a radio link failure.

3.4.3.3   Interference Control

Coordinated transmission and reception is a functionality provided in LTE advanced having the goals of interference reduction, higher data rates and higher cell edge throughput. It

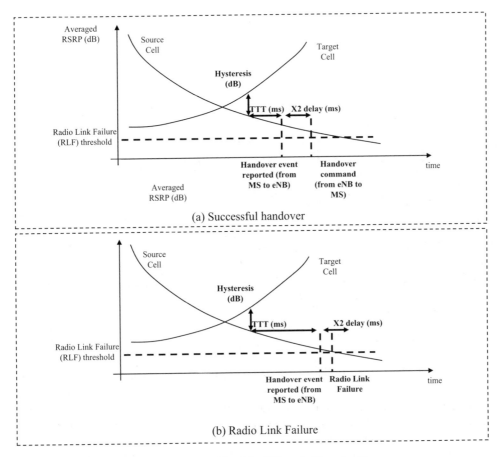

Figure 3.24   Example of the HO evaluation algorithm.

implies coordination among different transmission points. Interference control can be divided into [19]:

- Downlink interference coordination
- Uplink interference coordination

Three possible interference coordination scenarios are shown in Figure 3.25. In Figure 3.25(a), a case of uplink intercell interference coordination (ICIC) is presented: the scheduling of the same uplink resource block is coordinated between cells A and B, and is assigned to mobile stations experiencing low interference. The transmitted power is adjusted to minimize intercell interference.

Figure 3.25(b) shows downlink ICIC: in this case the mobile user is in the cell-edge area. Cells A and B coordinate the assignment of resource blocks to cell-edge users in order to maximize throughput and minimize interference. In the example, at time t_0 a resource block maximizing the mobile station (MS) throughput is assigned from cell A; at time t_1 a resource block that maximizes the MS throughput is assigned from cell B.

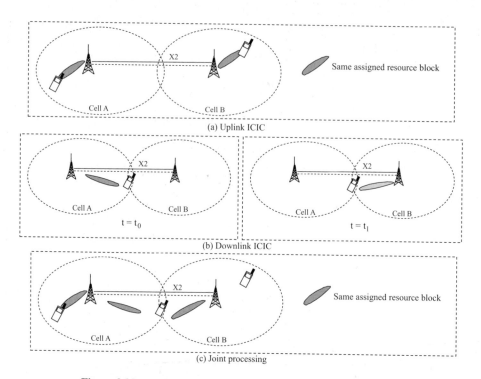

Figure 3.25   Example intercell interference coordination scenarios.

In Figure 3.25(c) the case of joint processing is shown: the same information is transmitted in downlink from cell A and cell B to the same mobile station and is jointly processed to improve the received signal quality.

Coordinated multipoint transmission and reception (CMTR) includes the following techniques:

- Coordinated radio resource management (RRM) and scheduling.
- Interference rejection combining (IRC): is a technique used in an antenna diversity system to suppress co-channel interference.
- Successive interference cancellation (SIC): is a multiuser detection technique in which the interfering signal of a particular user is cancelled after making a decision on that user's signal.

In LTE two indicators exchanged among adjacent eNBs are defined that aid uplink intercell interference coordination:

- Overload indicator (OI): characterizes the uplink interference for each resource block (RB) in a cell and is transmitted from an eNB to the neighbours. It indicates levels of high, medium and low interference.
- High interference indicator (HII): indicates the intention of an eNB to assign certain resource blocks to some cell-edge users.

Figure 3.26 shows an example of uplink intercell interference coordination.

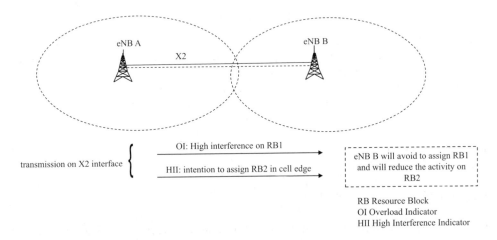

Figure 3.26   Example of uplink intercell interference coordination.

In the figure, eNB A transmits through the X2 interface to eNB2 that resource block 1 (RB1) experiences high interference (OI indicator) and that it has the intention to assign RB2 to cell-edge users (HII). eNB2 will avoid assigning RB1 and will reduce the activity on RB2.

In downlink, where the interference originates from the eNBs, intercell coordination implies the restrictions of the transmission power in some parts of the transmission bandwidth. To aid downlink intercell interference coordination, the following parameter exchanged through the X2 interface is defined:

- Relative narrowband transmission power indicator (RNTPI): indicates that the transmission power of a set of resource blocks is lower than a threshold. An eNB that receives this parameter will schedule its downlink transmission on the same set of resource blocks, thus allowing full frequency reuse.

Figure 3.27 shows an example of downlink intercell interference coordination.

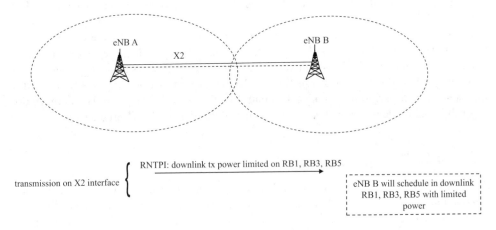

Figure 3.27   Example of downlink intercell interference coordination.

3.4.3.4   Coverage and Capacity Optimization

The coverage and capacity optimization function has the double objective of eliminating coverage holes and optimizing network capacity. The MS reports measurement results allowing the detection of coverage problems and the parameters involved in coverage and capacity optimization are adjusted to resolve coverage problems and optimize the capacity.

The involved parameters in the optimization, used to modify cell dimension and coverage, are [18]:

- Downlink transmit power
- Antenna tilt and azimuth

Typical coverage problems are [18]:

- Coverage hole: the received signal strength of the pilot channel is not sufficient to establish and/or maintain a connection. The network is aware of this type of failure because the mobile station experiences a radio link failure or a call drop. The connection is re-established in the same cell and HO parameter optimization does not solve the problem.
- Weak coverage: the received signal strength of the pilot channel is lower than an established threshold (connected to performance, such as that of the cell-edge bit rate).
- Pilot pollution: it occurs when overlapping causes too much interference.
- Overshoot coverage: it occurs when the coverage of a cell extends much further than it should. In this case the mobile station could experience high interference and/or call drops.
- Uplink/downlink coverage mismatch: in this case the downlink link budget and the uplink link budget are not matched. This causes problems in the connection establishment or maintenance. The network is aware of this type of failure because the mobile station experiences radio link failures or call drops.

Some of the parameters are involved in different self-optimization algorithms. In fact, self-optimization algorithms are mutually dependent and one control parameter is modified by different algorithms. This is why separate target optimization is not possible and algorithms must be able to jointly optimize multiple targets (i.e., interference, load balancing, HO, coverage).

Figure 3.28 shows the location of self-optimization logical functions and their connections for multiple target optimization. The operator through the network manager (NM) sets the policy control function (PCF) and monitors the optimization process through the monitoring functions (MFs). Policy control functions and monitoring functions are also performed at the element manager domain, under the control of the network manager. The optimization algorithms are implemented in the eNBs, connected through the X2 interface.

3.4.4   Energy Saving Management

Today, CO_2 emission reduction is a worldwide concern. The European Union has targeted a reduction of 20% for year 2020. Telecommunication networks consume a huge amount of energy for service delivery, and energy consumption is a hot topic for telecommunication equipment manufacturers. Radiomobile network nodes require a high amount of energy to

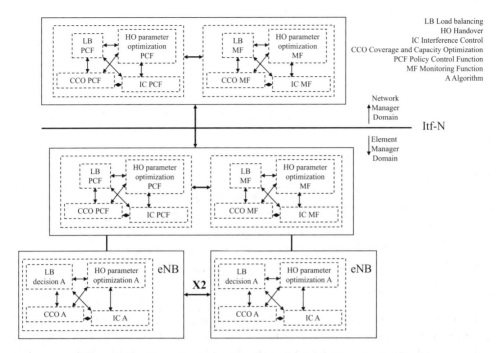

Figure 3.28   Example of the self-optimization logical function location.

provide service to the final users. Base stations, NBs and eNBs are the network nodes that consume most of the energy.

Energy saving (ES) has the objective of minimizing energy consumption through policies set by the operator [21]. Alternative energy solutions, like renewable energy sources, can also bring significant savings. The energy saving management function can be implemented at the network manager level (above the Itf-N interface) and at the element manager level (below the Itf-N interface). In both cases, the energy management is considered to be centralized. If ES is implemented at the network element level, ES is called distributed.

Energy saving involves not only LTE but also the other radio access technologies (RATs) and can be applied to single cells or network elements [4]. A cell or a network element can be, with respect to ES, in three states [21]:

- A notEnergySaving (NoES) state: the cell or the network element is working at full and no ES is activated.
- An EnergySaving state: the cell or network element is not working at full and ES is activated. In this case some functions are powered off.
- A compensatingForEnergySaving state: the network element compensates changes in ES states of other cells or network elements, for example to guarantee the same coverage when neighbour cells are switched off.

Figure 3.29 shows that the transition from the notEnergySaving state to the EnergySaving state is performed through ES activation, by switching off a cell or a network element. The

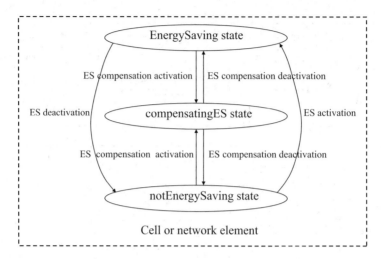

Figure 3.29 Energy saving states and transitions.

transition from the EnergySaving state to the notEnergySaving state is performed through ES deactivation, by using the cell or network element at full. ES compensation is a procedure that changes the cell/network element configuration with the goal of compensating changes in ES states of neighbour cells/network elements, in order to guarantee coverage and capacity. If in a cell or a network element ES is disabled, energy saving is forbidden.

If energy saving (ES) is centralized at the network manager level, the operator monitors the energy saving process and sets policies and conditions. Only when such policies and conditions are met are the cells instructed to move to the ES state. If ES is done at the level of the element manager or network element, the network element (i.e., eNB) moves autonomously to the ES state based on its own policies and conditions. The transition of some cells to the ES state results very often in the necessity to move the neighbour cells to the compensatingForEnergySaving state, in order to adjust coverage and meet the new target capacity.

In Figure 3.30 is shown an example of ES activation and ES compensation activation. When the network is at full traffic load, each eNB is in the NoES state. When the traffic decreases, eNBs B, C, D and E move to the ES state and are switched off. At the same time, eNB A moves to the ES compensation state and radio parameters are reconfigured in order to meet new traffic and coverage requirements. The coverage adjustment process among the eNBs in ES and the ES compensation state should be supported from the interference control and coverage and capacity optimization function. In order to optimize multiple targets, self-optimization functions like energy saving, interference control, and coverage and capacity optimization should be centralized, under operator control, at the network management level.

3.4.5 Self-Healing

Self-healing concerns the possibility to solve automatically the faults by triggering recovery actions [22]. The triggers for self-healing are the monitored alarms.

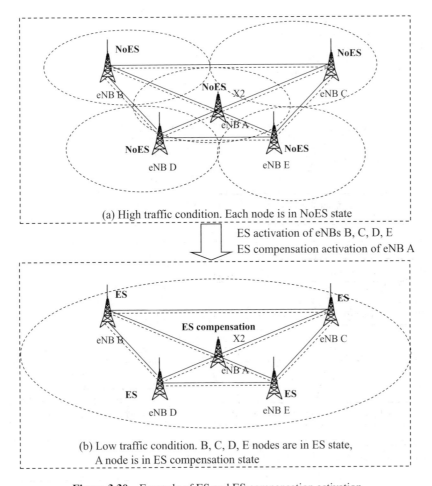

(a) High traffic condition. Each node is in NoES state

ES activation of eNBs B, C, D, E
ES compensation activation of eNB A

(b) Low traffic condition. B, C, D, E nodes are in ES state,
A node is in ES compensation state

Figure 3.30   Example of ES and ES compensation activation.

Monitored alarms and self-healing actions should be configurable by the operator. The self-healing function can be implemented at the network manager level (above the Itf-N interface) and at the element manager level (below the Itf-N interface). The self-healing function at the network manager level allows the operator to control execution of the self-healing process and to set the monitored alarms and self-healing actions. The self-healing process flow chart, split into the monitoring part and the healing process part [23], is shown in Figure 3.31.

In the monitoring part of the self-healing process, alarms are monitored and, if alarms that can be self-healed are found, an appropriate self-healing process is triggered. At the beginning of the self-healing process, fault related information is retrieved and analysed in order to identify the necessary recovery actions. At this point the recovery actions are performed. Then the self-healing results are evaluated and the process ends if the faults have been solved or if the stop condition of the self-healing process has been reached. Otherwise, the self-healing process is repeated.

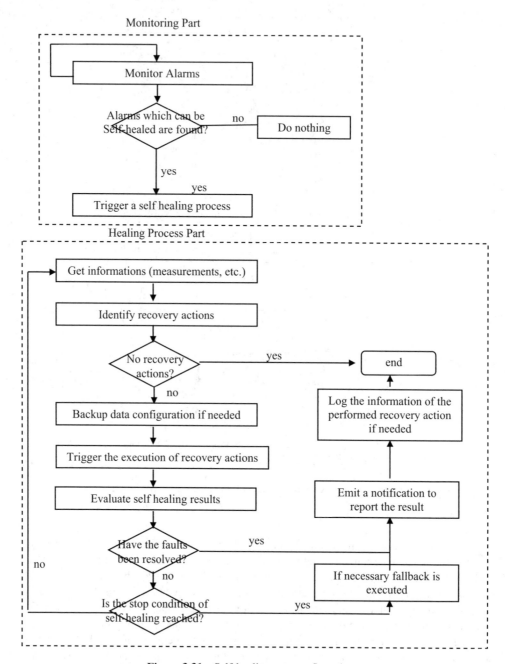

Figure 3.31   Self-healing process flow chart.

The faults can be software or hardware. In the case of hardware faults, the recovery actions can be:

- If the hardware is redundant, then the action is a changeover.
- If the hardware has no backup, then the actions can be different, depending on the faulty resource. For example, the faulty resource could be isolated, in order not to affect the other resources, or could be reset, or other reconfiguration actions can be performed.

In the case of software faults, the recovery actions can be [22]:

- System/process restart
- Reload of backup software
- Download of new software
- Data restoration
- Data reconfiguration
- Activation of a fallback software load (the previous software version is installed)

The reference model for self-healing is made of numerous functional blocks and is shown in Figure 3.32.

The self-healing monitoring and management function (SH_MMF), which controls the self-healing process, is split into two parts. The first part is implemented at the element manager level and is called the self-healing monitoring and management function element manager (SH_MMF_EM). It is able to get information and give feedbacks to all other functional blocs. The second part is implemented at the network manager level and is called the self-healing monitoring and management function network manager (SH_MMF_NM). It is connected to the SH_MMF_EM with the standard interface Itf-N and allows the operator to control execution

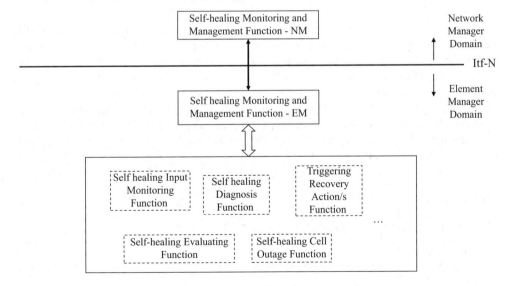

Figure 3.32   Self-healing reference model.

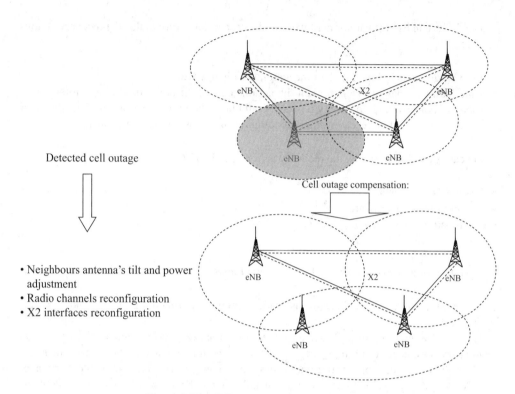

Detected cell outage

• Neighbours antenna's tilt and power adjustment
• Radio channels reconfiguration
• X2 interfaces reconfiguration

Figure 3.33 Cell outage and its compensation.

of the self-healing process [23]. At the network manager coordination among several eNBs must be implemented.

An example of self-healing involving other SON features and eNB coordination is cell outage. In this case, a cell goes out of service, cell outage is detected and the self-healing cell outage function (SH_CO_F) is triggered. The recovery action involves neighbour eNBs and other SON functions, like load balancing and coverage and capacity optimization. Radio parameters of the neighbour cells are reconfigured in order to avoid loss of service. For example, neighbour cell parameters like antenna tilt, power and channel configuration are modified in order to compensate for the loss of coverage and service. The neighbour list shall be updated and X2 interfaces shall be reconfigured. Figure 3.33 shows and example of cell outage and its compensation.

References

1. Overview of 3GPP release 8 V0.2.3 (2011-06).
2. Next Generation Mobile Networks Recommendation on SON and O&M Requirements, A Requirement Specification by the NGMN Alliance, December 2008.
3. Next Generation Mobile Networks Informative List of SON Use Cases, An Annex Deliverable by the NGMN Alliance, April 2007.
4. NGMN Top OPE Recommendations, A Deliverable by the NGMN Alliance, September 2010.

5. Lofrumento, G. (2011) Le applicazioni nel cloud: opportunità e prospettive. *Notiziario Tecnico Telecom Italia*, **20** (1).
6. Mell, P. and Grance, T. (2011) The NIST Definition of Cloud Computing, NIST Special Publication 800-145, September 2011.
7. 3GPP Technical Specification, TS 32.593, 3rd Generation Partnership Project; Technical Specification Group Services and System Aspects; Telecommunication Management; Home eNode B (HeNB) Operations, Administration, Maintenance and Provisioning (OAM&P); Procedure Flows for Type 1 Interface HeNB to HeNB Management System (HeMS), version V11.0.0, September 2011.
8. TR-069 Amendment 2, CPE WAN Management Protocol v1.1, Broadband Forum.
9. 3GPP Work Items on Self-Organizing Networks.
10. 3GPP Technical Specification, TS 32.500, 3rd Generation Partnership Project; Technical Specification Group Services and System Aspects; Telecommunication Management; Self-Organizing Networks (SON); Concepts and Requirements, version V0.3.1, July 2008.
11. 3GPP Technical Specification, TS 32.500, 3rd Generation Partnership Project; Technical Specification Group Services and System Aspects; Telecommunication Management; Self-Organizing Networks (SON); Concepts and Requirements, version V11.1.0, December 2011.
12. 3GPP Technical Specification, TS 32.501, 3rd Generation Partnership Project; Technical Specification Group Services and System Aspects; Telecommunication Management; Self Establishment of eNodeBs (SEe); Concepts and Requirements, version V0.2.0, July 2008.
13. 3GPP Technical Specification, TS 32.501, 3rd Generation Partnership Project; Technical Specification Group Services and System Aspects; Telecommunication Management; Self-Configuration of Network Elements; Concepts and Requirements, version V10.0.0, March 2011.
14. 3GPP Technical Report, TR 32.816, 3rd Generation Partnership Project; Technical Specification Group Services and System Aspects; Telecommunication Management; Study on Management of LTE and SAE, version 8.0.0, December 2008.
15. 3GPP Technical Specification, TS 25.484, 3rd Generation Partnership Project; Technical Specification Group Radio Access Network; Automatic Neighbour Relation (ANR) for UTRAN; Stage 2, version 10.1.1, June 2012.
16. 3GPP Technical Report, TR 36.300, 3rd Generation Partnership Project, Technical Specification Group Radio Access Network, Evolved Universal Terrestrial Radio Access (E-UTRA) and Evolved Universal Terrestrial Radio Access Network (E-UTRAN), Overall Description; Stage 2, version V11.2.0, June 2012.
17. 3GPP Technical Specification, TS 36.304, 3rd Generation Partnership Project; Technical Specification Group Radio Access Network; Evolved Universal Terrestrial Radio Access (E-UTRA); User Equipment (UE) Procedures in Idle Mode, version 11.1.0, September 2012.
18. 3GPP Technical Specification, TS 32.522, 3rd Generation Partnership Project; Technical Specification Group Services and System Aspects; Telecommunication Management; Self-Organizing Networks (SON) Policy Network Resource Model (NRM) Integration Reference Point (IRP); Information Service (IS), version 11.3.0, September 2012.
19. 3GPP Technical Specification, TS 32.521, 3rd Generation Partnership Project; Technical Specification Group Services and System Aspects; Telecommunication Management; Self-Organizing Networks (SON) Policy Network Resource Model (NRM) Integration Reference Point (IRP); Requirements, version 11.0.0, September 2012.
20. 3GPP Technical Specification, TS 36.331, 3rd Generation Partnership Project; Technical Specification Group Radio Access Network; Evolved Universal Terrestrial Radio Access (E-UTRA); Radio Resource Control (RRC); Protocol Specification, version 11.1.1, September 2012.
21. 3GPP Technical Specification, TS 32.551, 3rd Generation Partnership Project; Technical Specification Group Services and System Aspects; Telecommunication Management; Energy Saving Management (ESM); Concepts and Requirements, version 11.2.0, September 2012.
22. 3GPP Technical Report, TR 32.823, 3rd Generation Partnership Project; Technical Specification Group Services and System Aspects; Telecommunication Management; Self-Organizing Networks (SON); Study on Self-Healing, version 9.0.0, September 2009.
23. 3GPP Technical Specification, TS 32.541, 3rd Generation Partnership Project; Technical Specification Group Services and System Aspects; Telecommunication Management; Self-Organizing Networks (SON); Self-Healing Concepts and Requirements, version 11.0.0, September 2012.

4

IEEE 802.22: The First Standard Based on Cognitive Radio

4.1 White Spaces

The transition to digital television left some available spectrum in the VHF and UHF bands. Such an available spectrum is also called a digital dividend and a portion of it has been allocated to licensed mobile services. In Europe, the Middle East and Africa (EMEA) and Asia-Pacific (APAC) regions, the 790–862 MHz band was identified for International Mobile Telecommunication (IMT) services; in the Americas, except Brazil, the band of 698–806 MHz was identified for IMT services. Bangladesh, China, India, Korea, Japan, New Zealand and Singapore identified the band or portions of the band 698–790 MHz for IMT to align with the Americas. The spectrum holes left in the mentioned bands are called 'white spaces' and can be used for the deployment of new unlicensed wireless services.

The digital dividend bands are very interesting because of their radio propagation characteristics, like wide coverage and high building penetration. An idea of how such bands are valuable comes from the recent LTE spectrum allocations. The 800 MHz band has been paid by the operators much more than the frequencies above GHz.

The efficient use of white spaces is therefore very important for an optimal utilization of this valuable part of the radio spectrum. Because of that, a number of initiatives have emerged proposing that white spaces could be used for a range of applications, like the provision of wireless broadband services in rural areas. Other possible applications are machine-to-machine communications, high speed, short range data links, etc.

Television band devices (TVBDs), or white space devices (WSDs), operate in the digital dividend. They transmit without license as secondary users on the free spectrum holes of the digital dividend without causing interference to the primary, licensed, users of the spectrum (i.e., TV broadcasters, mobile operators). To do that, they must be aware of the radio context in that location (i.e. operating radio access networks, radio access networks capabilities, geo-location information, spectrum occupancy, etc.).

There are three methods used to find out the radio context, which can be used jointly: the cognitive pilot channel (CPC), spectrum sensing and geo-location database.

Reconfigurable Radio Systems: Network Architectures and Standards, First Edition. Maria Stella Iacobucci.
© 2013 John Wiley & Sons, Ltd. Published 2013 by John Wiley & Sons, Ltd.

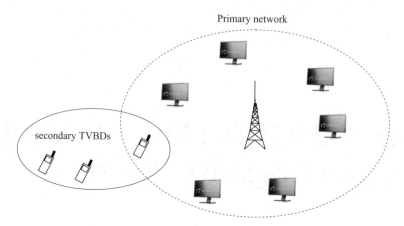

Figure 4.1   Scenario with licensed and unlicensed users.

The cognitive pilot channel is a signalling channel that broadcasts context information (i.e., available frequency bands, frequency bands usage, operating radio access networks, available services, etc.) in the location of the device. The cognitive pilot channel can be:

- In band: the signalling broadcast channel is transmitted on the same band of the data. For white spaces, the band is the digital dividend.
- Out band: the signalling broadcast channel is transmitted on a band outside the TV band.

Spectrum sensing is based on measurements from a television band device (TVBD) of a part of the spectrum in order to decide if it can transmit on that band [4]. The scenario, shown in Figure 4.1, provides the presence of primary users operating on licensed bands and secondary users with the possibility to transmit in an unlicensed manner. Spectrum sensing can be used to detect the presence of a primary user transmission in three dimensions: time, frequency and space [2].

The geo-detecting database is a complementary or alternative technique to spectrum sensing. It is a database containing information on the usage of the spectrum, which can be queried from television band devices (TVBDs) to find out if they are allowed to transmit over a certain part of the spectrum [3]. Figure 4.2 shows an example of a procedure for configuring and using a geo-detecting database.

When a radio cell of the primary network is switched on, the information about the new radio cell in terms of location and used frequencies are transmitted to the geo-location database. Before transmitting, a television band device (TVBD) queries the database to obtain the map of white spaces at that time in its location.

4.1.1   FCC Regulation

On 14 November 2008, the FCC allowed cognitive radio operation in white spaces, with spectrum sensing and a geo-detecting database to protect TV signals from the interference

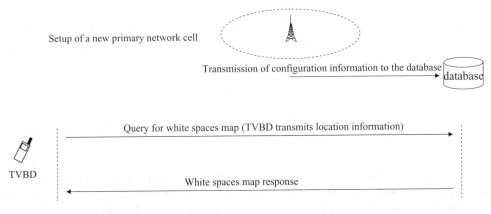

Setup of a new primary network cell

Transmission of configuration information to the database → database

Query for white spaces map (TVBD transmits location information)

TVBD

White spaces map response

Figure 4.2 Example of the procedure for configuring and using a geo-detecting database.

caused by cognitive devices transmissions [4]. Frequencies identified from FCC for white spaces range from 54 to 216 MHz (VHF) and from 470 to 698 MHz (UHF), as shown in Table 4.1. Channels 2 to 20 are reserved for fixed TVBDs only [5]; channel 37 (608–614 MHz) is reserved for radio astronomy, while channels 14 to 20 are shared with public safety [5].

Under the rules set in 2008 [4], cognitive devices operating in the digital dividend have to check on TV broadcasting activity every 60 seconds and have to perform a sensing test for wireless microphones and other low power signals every 30 seconds. If the device senses an activity on the channel, it stops the transmission. Device makers and other public interest groups strongly objected to the spectrum sensing requirement because it protects many wireless systems operating under the Federal Communication Commission (FCC) Part 15 rules. FCC Part 15 rules cover the regulations under which an intentional, unintentional or incidental radiator can be operated without an individual license. Such a radiator, as FCC states, 'must accept any interference received, including interference from other un-licensed devices'.

On 23 September 2010, a second memorandum opinion and order was released, where FCC eliminated the requirement of spectrum sensing, which can be included in the television band devices (TVBDs) on a voluntary and not a mandatory basis [6]. The cognitive pilot channel has not been considered for white spaces. The operation in TV white spaces is therefore allowed only by using a geo-location database. Two categories of unlicensed devices are

Table 4.1 TV channels and white space allocation for FCC

Channel number	Frequency band	
2, 3, 4	54–72 MHz	VHF
5, 6	76–88 MHz	
7, 8, 9, 10, 11, 12, 13	174–216 MHz	
14, 15, 16, 17, 18, 19, 20	470–512 MHz	UHF
21, 22, 23, . . . , 51	512–698 MHz	

permitted in the TV bands: fixed and portable TVBDs. Such devices must contact a geo-location database to obtain the list of available channels before operating. Fixed devices have a fixed location and are provided with geo-location capabilities. Two types of portable devices are allowed:

- *Mode II devices:* they have geo-location capabilities and are able to access the database.
- *Mode I devices:* they do not have geo-location capabilities and are not able to access the database; they receive a list of available channels from a mode II device or from a fixed device.

Then TVBDs will learn about white space availability at a certain time in a given location only from a central database of primary networks. The communication between TVBDs and the database must be secure, and all information required in the TV band database must be publicly available [6]. In the same second memorandum opinion and order, FCC also included protection criteria for incumbent authorized devices, technical rules for TVBDs, TV band database requirements and which channels can be used by TVBDs.

FCC defined in [6] the maximum operation power for fixed, personal and portable devices. A fixed device can operate with a maximum transmitter power output of 1 W and may use an antenna providing up to 6 dBi of gain [6]. Personal or portable devices are allowed to use up to 100 milliwatts e.i.r.p. (effective isotropic radiated power), with the exception of the channels adjacent to a TV station transmission, where they are limited to 40 mW e.i.r.p. [6]. TVBDs using only spectrum sensing are also allowed with a maximum transmission power of 50 mW e.i.r.p. The maximum height above average terrain (HAAT) for fixed devices was limited in [6] to 76 m.

On 5 April 2012, a third memorandum opinion and order was released [7], where the maximum height above average terrain for sites where fixed devices may operate was increased, the adjacent channel emission limits were modified to specify fixed rather than relative levels and the maximum power spectral density (PSD) for each category of TV band devices was slightly increased. In [7], the 76 m site HAAT limit for fixed devices has been removed and an HAAT of up to 250 m was permitted.

Figure 4.3 shows an example of fixed and portable TVBDs, with their power limits.

4.1.2   ECC Regulation

In Europe, the Electronic Communication Commission (ECC) identified frequencies for white spaces ranging from 470 to 790 MHz (UHF). The band from 790 to 862 MHz, used for TV channels before the transition to digital television (DTV), was identified for IMT services.

The ECC published on January 2011, within the European Conference of Postal and Telecommunications Administrations (CEPT), the ECC Report 159 [8], entitled 'Technical and Operational Requirements for the Possible Operation of Cognitive Radio Systems in the White Spaces of the Frequency Band 470–790 MHz'. The objective was to 'provide technical and operational requirements for cognitive radio systems (CRS) in the "white spaces" of the frequency band 470–790 MHz in order to ensure the protection of the incumbent radio services' [8].

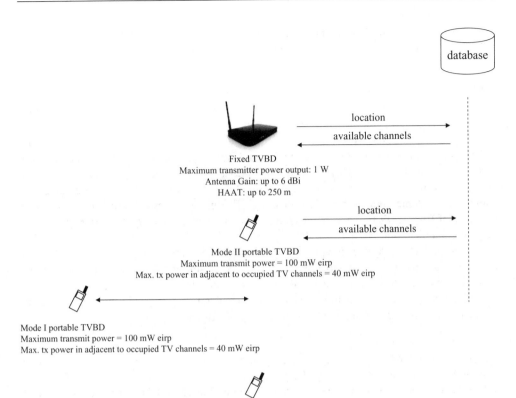

Figure 4.3   Example of fixed and portable TVBDs, with their power limits (under FCC rules).

The three techniques of the cognitive pilot channel, spectrum sensing and geo-detecting database have been studied. The conducted studies lead to the following considerations:

- The use of spectrum sensing only does not guarantee protection of the incumbent radio services.
- The use of a geo-location database seems to be the most probable choice to avoid interference; in that case spectrum sensing is an option in the implementation of wireless devices. The database stores information about the available white spaces in that location and the associated power limits.

Two ECC reports were published in January 2013:

- ECC Report 185, entitled 'Complementary Report to ECC Report 159: Further Definition of Technical and Operational Requirements for the Operation of White Space Devices in the Band 470–790 MHz' [9]

- ECC Report 186, entitled 'Technical and Operational Requirements for the Operation of White Space Devices under Geo-location Approach' [10]

In [9], some further studies to complement and enhance the results published in [8] are contained. In [10], complementary and enhanced results in relation to geo-location techniques acknowledging the one published in [8] are given.

In Report 159 it is stated that 'in some of . . . geo-location database usage models it may not be necessary for administrations to define, assume or mandate a fixed value for the maximum permitted e.i.r.p. for WSDs. However, administrations may still decide to assume or mandate maximum permitted e.i.r.p. of WSDs considering their usage and the DTT (digital terrestrial television) implementations they are protecting'. Examples of power limits of WSDs are in the range of 10–50 mW for short range communications and in the range of 1–10 W for longer range communications.

ECC Report 185 addresses the possibility to set up fixed maximum permitted power limits for WSDs. The geo-location database has the task to determine and communicate the maximum allowed transmission power to the TVBD, which is controlled by the database.

An example of a query algorithm to the geo-location database follows. To query the database, a TVBD is required to communicate the following information:

- Its location and location accuracy: the geographical coordinates and the uncertainty radius around them are transmitted to the database.
- The device type/class/model, device identifier, emission class, technology identifier: such parameters take into account the device capabilities and its technical characteristics.
- The expected area of operation (optional): the TVBD could estimate the possible area of operation and send it to the database as the radius of the circle having its position at the centre.

The database answers to the query with the following information:

- Available frequencies in the device's location, maximum transmit power: the database could take into account the received information about the device class (i.e., technical characteristics, interference characteristics) in the computation of available channels and tx (transmitter) power limits.
- Valid database to consult (optional): this optional information is useful for roaming purposes. In fact, if the user moves to an area managed by another database, it has to be informed about it.
- Time validity of the provided information: after that time, the TVBD must make a new query to the database even if it did not move.
- Requirement of sensing (optional): the database requests to the TVBD having sensing capabilities to perform sensing (i.e., to identify wireless microphone transmissions). After the sensing, the transmission to the database of the sensed available channels can be included [11].

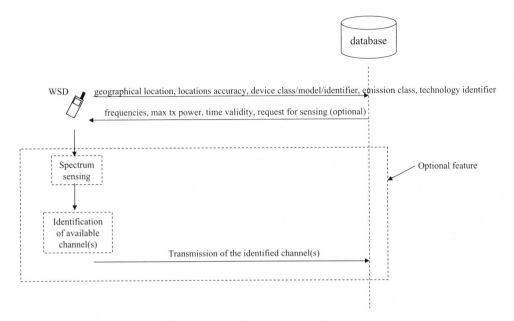

Figure 4.4   Example of the query algorithm to the geo-location database.

Figure 4.4 shows an example of the query algorithm to the geo-location database.

If a TVBD is not able to query the database, it can act as a slave of a master TVBD having the query capability [8, 10]. The slave TVBD communicates its location, location accuracy and device type to the master. The master queries the database and communicates the available frequencies to the slave. The master is responsible for assigning to the slave the communication resources (frequencies and maximum transmission power) in order to guarantee interference protection to primary users. As an optional feature, the master can ask a slave TVBD to perform sensing on the available channels [11]. An example of the master–slave algorithm is shown in Figure 4.5.

4.2   IEEE 802.22

The IEEE 802 group specifies PHY and MAC layers of the protocol stack for different fixed and wireless networks. A resume of the most important IEEE standards for wireless networks is shown in Figure 4.6.

IEEE 802.11 is a standard for wireless local area networks (WLANs). IEEE 802.15 is for wireless personal area networks (WPANs) and includes Bluetooth as IEEE 802.15.1, the standard IEEE 802.15.4 for low-rate WPANs and the standard IEEE 802.15.3 for high-rate WPANs.

IEEE 802.16 is for wireless metropolitan area networks (WMANs) and includes fixed WiMAX as the IEEE 802.16d standard, mobile WiMAX as the IEEE 802.16e standard and IEEE 802.16m, which satisfies the ITU requirements for a fourth generation broadband mobile communications system.

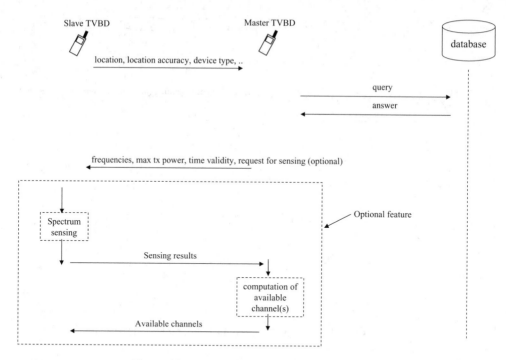

Figure 4.5   Example of the master–slave algorithm.

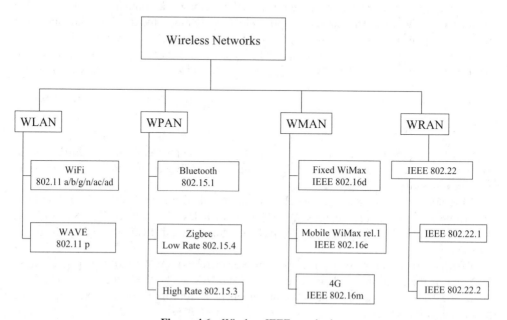

Figure 4.6   Wireless IEEE standards.

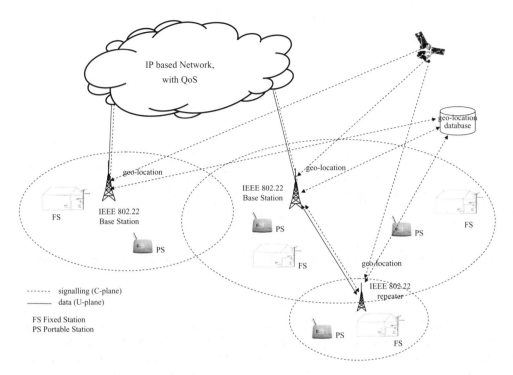

Figure 4.7   Example of the IEEE 802.22 network.

IEEE 802.22 is the first standard using white spaces based on a cognitive radio, published on 1 July 2011 [12]. It is a standard for wireless regional area networks (WRANs), which enables broadband wireless access, especially in rural areas [13]. Inside IEEE 802.22, two working groups have been created:

- IEEE 802.22.1, to define a standard for cognitive radio devices to guarantee the protection of licensed devices operating at low power. The standard was published on 1 November 2010.
- IEEE 802.22.2, with the aim of defining recommended practices for installation and deployment of IEEE 802.22.

An example of the IEEE 802.22 network is shown in Figure 4.7.

The network topology is point-to-multipoint, with IEEE 802.22 base stations (BSs) connected to the Internet backbone. Base stations acting as repeaters can be included in the network architecture. The BS realizes the radio coverage and provides a connectivity service up to 512 fixed or portable stations. Each BS is provided with geo-location capabilities and communicates with the geo-location database. In the example of Figure 4.7, the geo-location is satellite-based. The supported range is typically up to 30 km, depending on the effective isotropic radiated power (e.i.r.p.) and on the antenna eight. However, the MAC of the IEEE 802.22 system is also able to handle users away from the BS up to 100 km [12]. Figure 4.8 shows an IEEE 802.22 base station connected to the database providing service to fixed and portable devices.

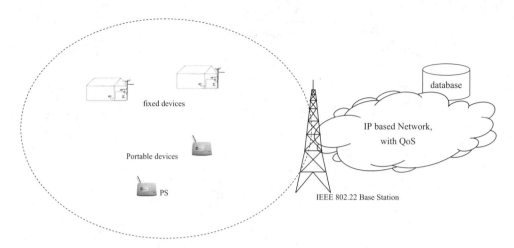

Figure 4.8 IEEE 802.22 base station connected to the database providing service to fixed and portable devices.

The standard includes cognitive radio techniques to minimize the interference to the primary users, like spectrum sensing and geo-location capabilities, provides access to a database and introduces the IEEE 802.22.1 wireless beacons. The IEEE 802.22.1 introduces a digitally modulated wireless beacon signal, which can be used to indicate the presence of nearby wireless microphone operation to unlicensed devices operating in the TV white space.

4.2.1   IEEE 802.22 Architecture

In the IEEE 802.22 architecture, a base station (BS) realizes the radio coverage and transmits with up to 512 terminals. An IEEE 802.22 network may consist of more BSs connected through an IP-based network and managed by entities implemented in a centralized or distributed manner.

The IEEE 802.22 management reference architecture is represented in Figure 4.9 and includes, apart from the base stations and the database, a network control system (NCS) and a network management system (NMS). In particular, the customer premise equipments (CPEs) and the base stations (BSs) are nodes managed from the NCS and NMS. All the nodes are able to access a database service, which is aware of the context in that location, such as usage of the spectrum with related policies, etc. The database service can be queried from television band devices (TVBDs) to know whether they are allowed to transmit over a certain part of the spectrum. The communication between the database service and the IEEE 802.22 network nodes is secure. The base stations, the network control system and the network management system are connected through an IP-based transport network.

Figure 4.10 shows the IEEE 802.22 network reference model. The customer premise equipments (CPEs) and the base stations (BSs) are nodes managed from a network control and management system (NCMS), implemented in the network and user side. In the figure, multiple CPEs are connected to the same base station (BS).

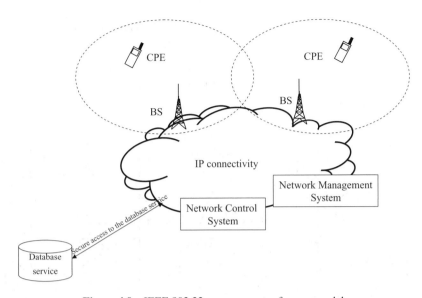

Figure 4.9   IEEE 802.22 management reference model.

The CPE communicates to the BS through the U standard interface. The managed objects like BSs and CPEs support a management protocol such as the simple network management protocol (SNMP). SNMP is a network protocol defined by the IETF (Internet Engineering Task Force), which operates at the application layer and allows the configuration, management and supervision of devices connected through a network. The interfaces between NCMS and

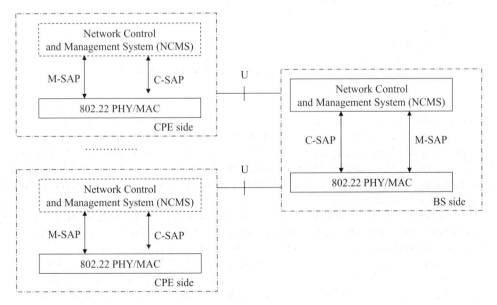

Figure 4.10   IEEE 802.22 network reference model.

802.22 Base Station

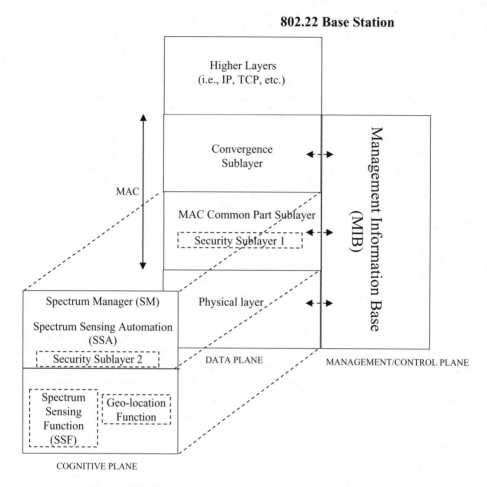

Figure 4.11   Protocol reference model (PRM) of IEEE 802.22 BS.

managed nodes are in the figure as M-SAP and C-SAP, which provide NCMS access to the control plane and management plane functions from the upper layers. M-SAP is used for low time sensitive management plane primitives, like system configuration, statistics, sensing reporting, geo-location reporting, etc. C-SAP is used for high time sensitive control plane primitives, like subscriber and session management, security context management, radio resource management, etc.

In order to handle the operation in the vacant channels of the TV bands without causing interference to the incumbents, the BS uses cognitive radio capabilities like geo-location, access to a geo-location database, spectrum sensing, spectrum management, etc.

The protocol reference mode (PRM) proposed in IEEE 802.22 is shown in Figure 4.11 for the BS side and in Figure 4.12 for the CPE side [12]. The PRM adds to the usual data and management/control plane functionalities a cognitive plane, supporting cognitive radio capabilities.

802.22 CPE

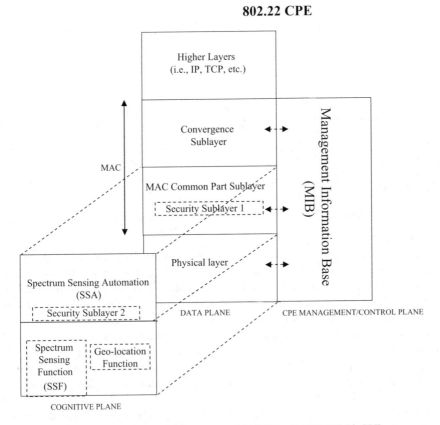

Figure 4.12   Protocol reference model (PRM) of IEEE 802.22 CPE.

4.2.1.1   IEEE 802.22 Data Plane

The data plane carries, other than data, management and control information. It includes the physical layer (PHY), the MAC layer and the convergence sublayer. Service access points (SAPs) are in between the layers because they are the points of access to a service offered from one layer to the other. In other words, the SAPs define an interface of a set of primitives used to exchange information between the layers.

MAC Common Part Sublayer
The MAC layer includes the common part sublayer (CPS) and the security sublayer 1, which provides mechanisms for authentication, secure key exchange, encryption, etc. The common part sublayer is based on a synchronous structure of superframes.

Two modes of operation are provided: normal mode and self-coexistence mode. In normal mode, a cell transmits on one TV channel and operates on all the frames of the superframe; in self-coexistence mode, more cells share the same channel and each cell operates only on one or several frames. The frames can be structured in normal or self-coexistence mode. The two modes are represented in Figures 4.13 and 4.14.

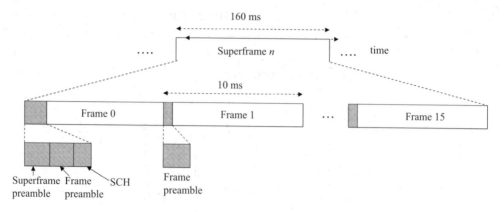

Figure 4.13 Example of a superframe structure in normal mode.

In both modes, each superframe lasts 160 ms and contains 16 frames of 10 ms each. The principal difference is that in normal mode only a wireless RAN transmits on the superframe; in self-coexistence mode a superframe is shared among different wireless RANs. In normal mode (Figure 4.13) the superframe preamble and superframe control header (SCH) are transmitted only at the beginning of a superframe; in self-coexistence mode (Figure 4.14) the superframe preamble and superframe control header (SCH) are not transmitted only at the beginning of the superframe because each wireless RAN transmits all the fields (superframe preamble, frame preamble and SCH) at the beginning of the first frame allocated to it. Figures 4.15 and 4.16 show respectively an example of normal mode and self-coexistence modes of operation.

In Figure 4.15, BS1 operates in the TV channel ch1 and BS2 operates in the TV channel ch2 (normal mode). In this case, each BS transmits its superframe preamble, frame preamble and SCH at the beginning of each superframe. In Figure 4.16, BS1 and BS2 share the same

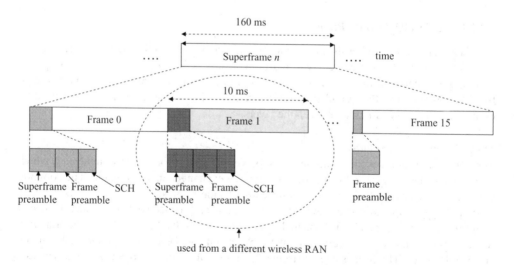

Figure 4.14 Example of a superframe structure in self-coexistence mode.

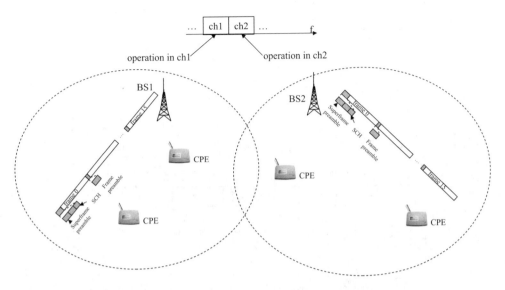

Figure 4.15　Example of normal mode of operation.

TV channel ch1. In this case, they operate in self-coexistence mode and a unique superframe is shared between the two BSs. In the figure, frames from zero to eight are allocated to BS1; frames from nine to fifteen are allocated to BS2. Each BS, at the beginning of its allocated period (frames), transmits its superframe preamble, frame preamble and SCH. The superframe and frame preambles are used from a CPE to synchronize with the superframe and with each frame. A CPE must receive the subframe control header (SCH) to associate and establish a communication with a BS.

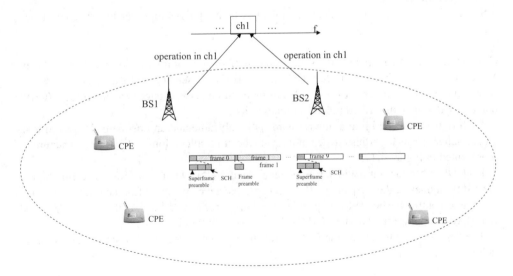

Figure 4.16　Example of self-coexistence mode of operation.

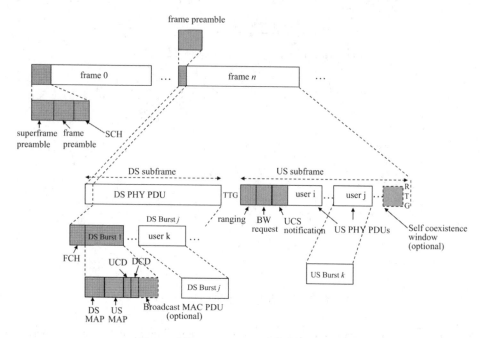

Figure 4.17   IEEE 802.22 frame structure.

The SCH carries information like:

- BS_ID: is the MAC address that identifies the BS transmitting in the SCH.
- Frame allocation map: indicates which frames in the subframe are allocated to the BS and which is transmitting in the SCH.
- Self-coexistence capability indicator: indicates whether the BS supports the self-coexistence mode.

Other fields in the SCH include the superframe number and parameters for the physical layer. Moreover, the BS schedules in the SCH quiet periods for sensing and supports coexistence with incumbents and other IEEE 802.22 cells by including coexistence beacon protocol (CBP) bursts. CBP will be described in later in this section.

Each frame is divided into a downstream (DS) subframe and an upstream (US) subframe, whose lengths can be adjusted depending on the time allocations. The frame structure is represented in Figure 4.17.

Each DL subframe starts with a frame preamble, used for synchronization and estimation of the radio channel, of the frequency offset and of the received power. After the frame preamble, the frame control header (FCH) specifies the burst profile and the length of DS-MAP and US-MAP (mapping) fields. The FCH field is followed by DS bursts.

The first DS burst following the FCH is transmitted in broadcast and contains the following fields:

- DS-MAP (downstream MAP): it is the first MAC-PDU of the first DS burst and defines the structure of the DS subframe.

- US-MAP (upstream MAP): it defines the structure of the US subframe.
- UCD (upstream channel descriptor): it is transmitted in broadcast and defines the characteristics of the upstream physical channels.
- DCD (downstream channel descriptor): it is transmitted in broadcast and defines the characteristics of the downstream physical channels.
- Broadcast MAC PDU (optional): it is optional and carries other broadcast information.

The other DS bursts of the DS subframe are allocated to the users. They include several MAC protocol data units (PDUs) and a padding field (PAD) if required. The US subframe is separated from the DS subframe by a guard time called Transmit/receive Transition Gap (TTG).

The US subframe starts with the following fields:

- Ranging slots: used to perform initial and periodic ranging.
- BW request slots: used from the CPEs to request resources to the BS.
- Urgent coexistence situation (UCS) notification slots: used by CPEs to notify the BS that an incumbent has been detected on the same channel.

At the end of the upstream frame, an optional self-coexistence window is used for the coexistence beacon protocol (CBP), which is based on the transmission of CBP packets carrying information about the cell and other coexistence mechanisms. Each MAC PDU contains a MAC header, a MAC payload and a CRC. The structure of DS and US bursts, and of the MAC PDUs is represented in Figure 4.18.

Figures 4.13, 4.14, 4.17 and 4.18 show the bursts of the frame and the superframe only in the time domain, following a time multiplexing principle. The information is actually transmitted in the two-dimensional time/frequency domain, with 2048 orthogonal frequency division multiple (OFDM) subcarriers, as exemplified in Figure 4.19. A burst is a two-dimensional segment of OFDM subchannels (in the frequency domain) and symbols (in the time domain). A subchannel is the basic unit for subcarrier allocation in upstream and downstream and is composed of 28 subcarriers.

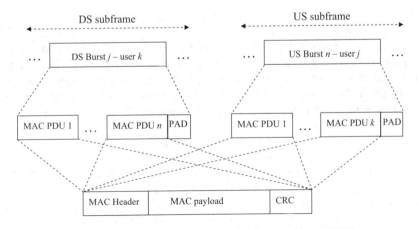

Figure 4.18 Structure of DS burst, US burst and MAC PDUs.

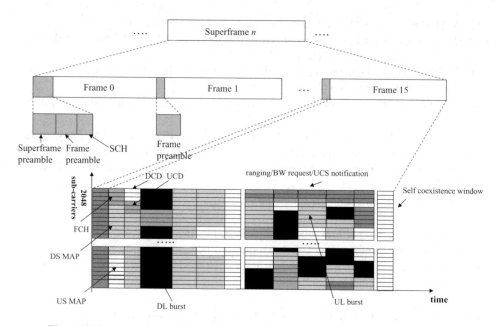

Figure 4.19   Example of the time/frequency structure of an IEEE 802.22 MAC frame.

IEEE 802.22 Physical Layer

The IEEE 802.22 physical layer supports the time division duplex (TDD) and is based on orthogonal frequency division multiple access (OFDMA) for both downlink and uplink. OFDMA is a multiple access technique based on OFDM modulation, a particular case of multicarrier transmission. In OFDM a high bit rate bit stream is split into multiple streams at low bit rates, each of which is QAM-modulated on a separated subcarrier.

OFDM subcarriers (unlike FDM subcarriers) are partially overlapping, thus allowing a considerable saving in terms of bandwidth. OFDM modulator and demodulator schemes have been introduced in Chapter 1, Section 1.2.6.2. Figure 4.20 shows the equivalent OFDM baseband transmission scheme with a cyclic prefix (CP).

At the transmission side, N modulation symbols X_i (i.e., QAM symbols) are put as input to the IFFT block to generate the OFDM symbol $\{x_1 \ x_2 \ldots x_N\}$. The presence of a channel with memory generates inter-OFDM symbol interference. In order to eliminate such interference, a cyclic prefix of v symbols is added to the transmitted sequence of $\{x_i\}$. The periodicity of the input sequence is simulated by replicating the last v samples of the sequence and placing them in the head. The CP of length v is inserted in a channel with memory v to eliminate inter-OFDM symbol interference. Then, the symbol time is:

$$T_{SYM} = T_{FFT} + T_{CP} \tag{4.1}$$

where T_{FFT} is equal to the inverse of carrier spacing Δf. T_{CP} for an IEEE 802.22 WRAN can be one of the following values: $T_{CP} = T_{FFT}/32$; $T_{CP} = T_{FFT}/16$; $T_{CP} = T_{FFT}/8$; $T_{CP} = T_{FFT}/4$.

IEEE 802.22 uses TV channels of 6, 7 or 8 MHz with a number of subcarriers equal to 2048 and supported modulations 4-, 16- and 64-QAM. Table 4.2 shows the principal system parameters.

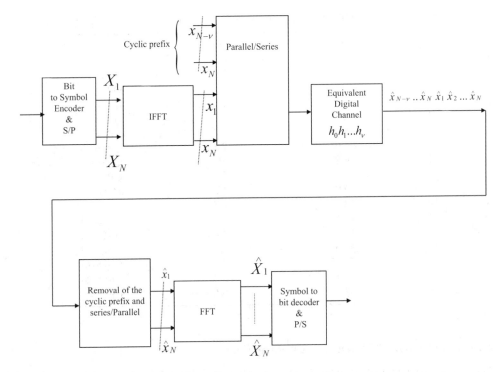

Figure 4.20 Equivalent OFDM baseband transmission scheme with a cyclic prefix.

Link adaptation adapts the choice of the modulation and coding to the received channel quality [14]. Table 4.3 shows the IEEE 802.22 modulation and coding schemes for data communications when the used bandwidth is of 6 MHz and $T_{CP} = T_{FFT}/16$. For bands of 7 or 8 MHz, the modulation and coding schemes are the same, but sampling frequency, carrier spacing and symbol duration scale with the bandwidth.

Physical mode 1 is used for CDMA opportunist bursts; physical mode 2 is used for SCH packet transmission; physical mode 3 is used for CBP transmission; physical mode 4 is used for SCH transmission. CDMA bursts are used for initial ranging and for terrestrial-based geo-location; CBP is a protocol based on the transmission of packets carrying information able to help the coexisting IEEE 802.22 cells to manage the shared spectrum. The CBP protocol is described in Section 4.2.1.4 of this chapter. Physical modes 1, 2, 3 and 4 do not have the corresponding bit rates because they are used for signalling.

Table 4.2 IEEE 802.22 principal system parameters

System parameter	Value
Frequency range	54–862 MHz
Bandwidth	6, 7 or 8 MHz
Multiple access	OFDMA
Number of subcarriers	2048
Duplexing	TDD

Table 4.3 IEEE 802.22 modulation and coding schemes for data transmission

PHY mode	PHY bit rate (Mbps)	Coding rate	Modulation
1		Uncoded	BPSK
2		1/2 repetition: 4	QPSK
3		1/2 repetition: 3	QPSK
4		1/2 repetition: 2	QPSK
5	4.54	1/2	QPSK
6	6.05	2/3	QPSK
7	6.81	3/4	QPSK
8	7.56	5/6	QPSK
9	9.08	1/2	16-QAM
10	12.10	2/3	16-QAM
11	13.61	3/4	16-QAM
12	15.13	5/6	16-QAM
13	13.61	1/2	64-QAM
14	18.15	2/3	64-QAM
15	10.42	3/4	64-QAM
16	22.69	5/6	64-QAM

4.2.1.2   IEEE 802.22 Management/Control Plane

The management control plane handles the management information base (MIB). It is a virtual database used for managing network entities. The protocol adopted for the communication with the MIB database and configuration, management and supervision of network entities (i.e., BSs, CPEs, routers) is the simple network management protocol (SNMP).

The SNMP is a network protocol defined by the IETF (Internet Engineering Task Force) in RFC (request for comments) 1155 [15] operating at the application layer and allows the configuration, management and supervision of devices connected through a network. Other RFCs related to the SNMP are RFC 1213 [16] and 1157 [17].

4.2.1.3   IEEE 802.22 Cognitive Plane

The cognitive plane includes spectrum sensing (SS) and geo-location (GL) at the physical layer and spectrum manager (SM) and spectrum sensing automation (SSA) at the MAC layer. The cognitive plane also includes a dedicated security sublayer 2.

Spectrum Sensing
Spectrum sensing is implemented at the physical layer of the cognitive plane. It is performed from the CPE to detect the available BSs in the area and the presence of incumbents. Spectrum sensing is used from BSs and CPEs to sense the available spectrum, select the best available channel, use the selected channel in coordination with other users or base stations, and release it when a licensed user is detected. It helps to solve an incumbent's detection and protection.

Spectrum sensing principles and techniques are described in Section 2.2.1 of this book.

Geo-location

Cognitive radio uses unoccupied bands for data transmission. Because the unused bands vary with position, the location information is used to help spectrum management algorithms.

Geo-location techniques can be satellite-based or terrestrial-based depending on whether the anchor points are satellites or terrestrial network nodes (i.e., base stations). An example of satellite-based geo-location is the global positioning system (GPS). A GPS receiver calculates its position by synchronizing to the signal sent by GPS satellites. Each GPS satellite continually transmits messages including the time the message is transmitted and the satellite position at that time. The position information is used from the receiver as a known point for estimating the distance to each satellite to apply localization algorithms based on trilateration. A review of trilateration and other geo-location techniques is provided in Section 2.2.1.4 of this book.

IEEE 802.22 puts satellite-based geo-location as mandatory and also supports terrestrially-based geo-location capabilities. According to IEEE 802.22, each BS and each CPE uses its satellite-based geo-location capability at the physical layer to determine the latitude and the longitude of its transmitting antenna [12]. Each CPE then provides its coordinates (i.e., latitude and longitude) to the BS during the registration procedure through an NMEA (National Marine Electronic Association) string. Figure 4.21 shows an example of the registration procedure.

Registration request (REG-REQ) and registration response (REG-RSP) messages may carry information elements (IEs) supporting the registration procedure. In the REG-REQ message an NMEA location string is transmitted together with other information like the supported IP version, the CPE capabilities, the support of ARQ, the CPE antenna parameters, etc. In the REG-RSP, the NMEA location is acknowledged and the configuration parameters are given to the CPE.

A WRAN can be provided of a location database service that knows the available channels related to the locations, assigns one or more channels to the CPEs or BSs, together with the maximum allowed power and time validity. The CPEs send to the database the location coordinates to update its information.

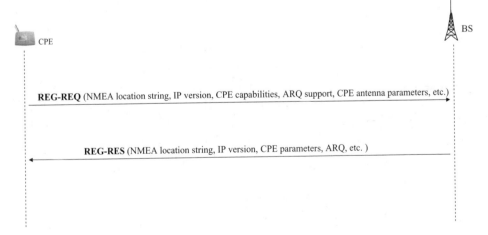

Figure 4.21   Example of the registration procedure.

Spectrum Sensing Automation

Spectrum sensing automation (SSA) is implemented at the MAC layer of the cognitive plane in BSs and CPEs. The SSA interfaces the spectrum sensing function (SSF) and executes the commands received from the spectrum manager (SM) to enable spectrum sensing. The SSA performs in-band and out-of-band sensing. Sensing is in-band when the measurements are performed in the operating channel and the first adjacent channels (N and N ± 1), and is out-of-band in all other cases.

For in-band sensing, the spectrum manager (SM) at the BS side schedules quiet periods (QPs) in the in-band spectrum and signals the scheduled QPs to the CPEs through the superframe control header (SCH) of the MAC frame. The CPEs use the scheduled periods to perform in-band spectrum sensing. If an incumbent is discovered, then the CPE sends a notification to the BS using the UCS (urgent coexistence situation) flag in the header of the next upstream PDU or through a UCS notification in an opportunistic notification window in the uplink MAC frame. An example of the in-band spectrum sensing procedure is shown in Figure 4.22.

Out-of-band sensing is performed while the CPE is in an idle state. The CPE reports to the BS the sensing results. The BS is then able to update the channel status in its channel list. Out-of band sensing can also be expressly requested from the BS through an out-of-band sensing results request message.

The BS can also request the CPEs to perform measurements on certain channels through MAC measurement messages, such as a single measurement request or the bulk measurement request (BLM_REQ). If the measurement request is through BLM_REQ, more than one measurement in different channels can be requested from the BS via a unicast, multicast or broadcast message to one or multiple CPEs. In the message timing parameters are also included to indicate the time intervals for measurements. The bulk measurement response

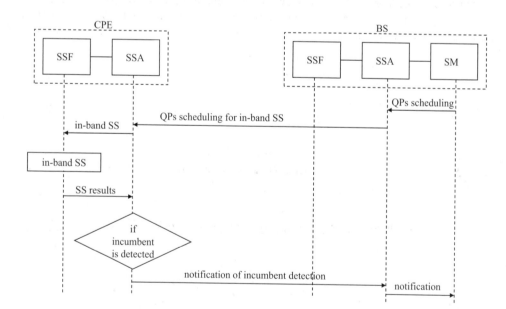

Figure 4.22   Example of in-band spectrum sensing procedure.

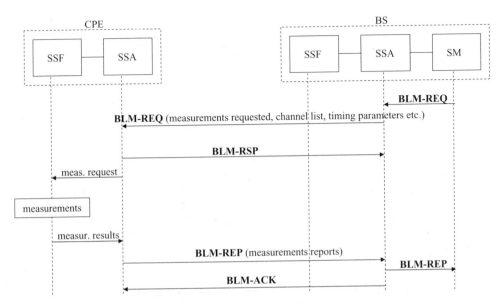

Figure 4.23   Example of measurement management procedure.

(BLM_RSP) message is used to acknowledge the reception of BLM_REQ. The CPE reports its measurements through the BLM_REP (bulk measurement report) message containing the measurement reports. Finally, BLM_REP is acknowledged through the BLM_ACK message. An example of the measurement management procedure is shown in Figure 4.23.

In general, SSA controls the spectrum sensing function in the following cases. At the BS side:

- When the BS is turned on
- During out-of-band sensing when the BS is not transmitting

At the CPE side:

- When the CPE is turned on
- During quiet periods (QPs) scheduled from the BS spectrum manager for in-band sensing
- During the CPE idle periods
- When the CPE loses contact with the BS

Database Service

IEEE 802.22 assumes a point-to-multipoint network architecture where a base station (BS) controls the radio parameters of its associated CPEs. The BSs are connected to a database service containing information on the usage of spectrum and the available channels. The interface to the database takes place between the IEEE 802.22 base station (BS) and the geo-location database. Figure 4.24 shows an example of an IEEE 802.22 WRAN with database service.

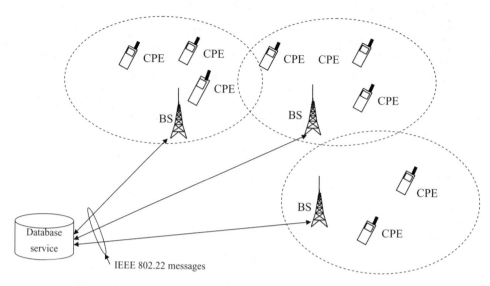

Figure 4.24 Example of an IEEE 802.22 WRAN with database service.

At the beginning, each BS will register to the database as a fixed device. Then the BSs will also register to the database their associated CPEs with their geographical location, CPEs identifier, etc. The BSs will also periodically query the database using the M-DB-AVAILABLE-CHANNEL-REQUEST message to obtain information about the channels in use. The database will also send to the BSs updates through push messages.

Database service primitives have been defined in the IEEE 802.22 standard for the communication between a BS and the geo-location database. The database service primitives are [12]:

- M-DB-AVAILABLE-REQUEST: the BS verifies if its connection to the database is available.
- M-DB-AVAILABLE-CONFIRM: the database confirms the availability of the connection to the BS.
- M-DEVICE-ENLISTMENT-REQUEST: the BS enlists to the database a device that has joined its network.
- M-DEVICE-ENLISTMENT-CONFIRM: the database confirms to the BS the device enlistment.
- M-DB-AVAILABLE-CHANNEL-REQUEST: the BS requests to the database the list of available channels with the maximum allowed e.i.r.p. for an enlisted device.
- M-DB-AVAILABLE-CHANNEL-INDICATION: the database indicates to the BS the available channels with their maximum allowed power for that device.
- M-DB-DELIST-REQUEST: the BS requests to the database to delist a device that is no longer in its network.
- M-DB-DELIST-CONFIRM: the database confirms the device that the device has been delisted.

Figure 4.25 shows an example of communication between a BS and the database.

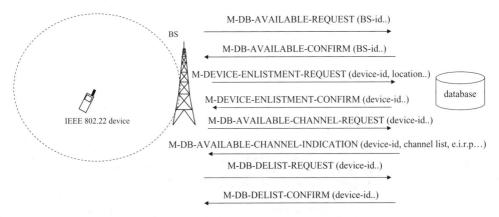

Figure 4.25   Example of communication between a BS and the database.

Spectrum Manager

The spectrum manager (SM) is implemented at the MAC layer of the cognitive plane at the BS side. It is a very important part of the base station (BS) because it is responsible for all the tasks related to the spectrum management, like maintaining spectrum availability information, channel selection and management, scheduling the spectrum sensing operation, etc. It guarantees the enforcement of spectrum policies, accesses the database service and assures incumbent protection. It uses the inputs coming from the geo-location and spectrum sensing function at the physical layer, the output of spectrum sensing automation and the information coming from the database service (i.e., spectrum policies, list of available channels, etc.) in order to decide which is the optimum channel for the WRAN cell and the power limits for the CPEs. An example of information exchange among the spectrum manager and other IEEE 802.22 entities is shown in Figure 4.26.

The SM communicates with the database to receive information on the availability of channels from the database service and classifies the channels as available or unavailable. The available channels can be classified as:

- Protected: the channel is in use (by an incumbent or an 802.22 WRAN).
- Unclassified: the channel has not yet been sensed.
- Disallowed: the channel is not allowed because of regulatory constraints.
- Operating: the channel is in operation for communication between a BS and CPEs in a cell.
- Backup: the channel has been released and can become the operating channel if the cell needs to move to another channel.
- Candidate: the channel is a candidate to become a backup channel.

Figure 4.27 shows the channel classification flow diagram.

When the channel classification procedure starts, the spectrum manager (SM) verifies that the database service exists to obtain the list of available channels from the database. For each available channel, spectrum sensing is performed. If the available channel cannot be sensed, it is unclassified; otherwise the spectrum manager takes a decision on the sensed channel and classifies it.

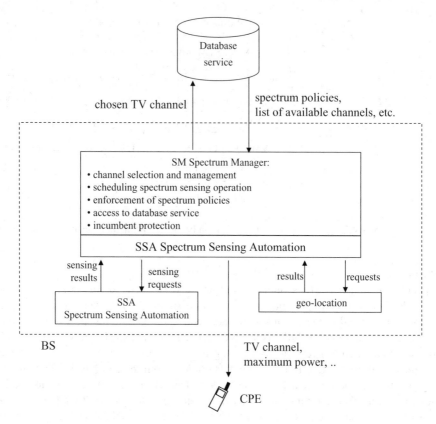

Figure 4.26 Example of information exchange among the spectrum manager and other IEEE 802.22 entities.

4.2.1.4 Coexistence

IEEE 802.22 provides two modes of operation: normal mode and self-coexistence mode. In normal mode, a cell transmits on one TV channel and operates on all the frames of the superframe; in self-coexistence mode, more cells share the same channel and each cell operates on one or several frames of the superframe [18]. Coexistence is managed at the MAC layer through the coexistence beacon protocol (CBP), which is a transport mechanism for inter-WRAN communications and self-coexistence, as shown in Figure 4.28.

Figure 4.28 shows that the CBP is a transport mechanism for two MAC self-coexistence schemes:

- Spectrum etiquette: it is used if there are enough available channels and orthogonal operating and backup channels can be found for different WRAN cells, meaning that the WRAN cells works in normal mode of operation.
- Frame-based on-demand spectrum contention: it is used if there are not enough available channels and different WRAN cells must share the same channel, thus working in self-coexistence mode of operation.

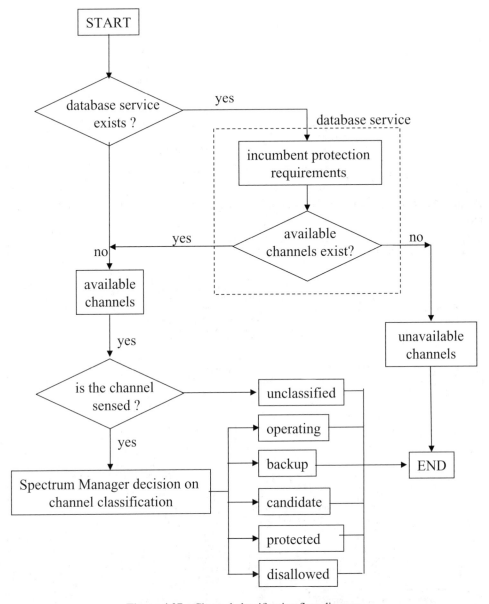

Figure 4.27   Channel classification flow diagram.

The execution flow for the inter-BS self-coexistence mechanism is shown in Figure 4.29. When the BS switches on, it performs network discovery operations, that is it searches the channels used from the neighbouring WRAN cells, the self-coexistence window reservations of the neighbouring WRAN cells and other information. Then the BS runs a channel acquisition algorithm based on spectrum etiquette. If the algorithm is successful, then enough available channels have been found and the BS runs in normal mode.

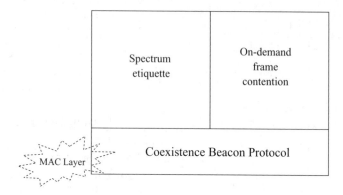

Figure 4.28 Coexistence beacon protocol (CBP) as a transport mechanism for inter-WRAN communications and self-coexistence.

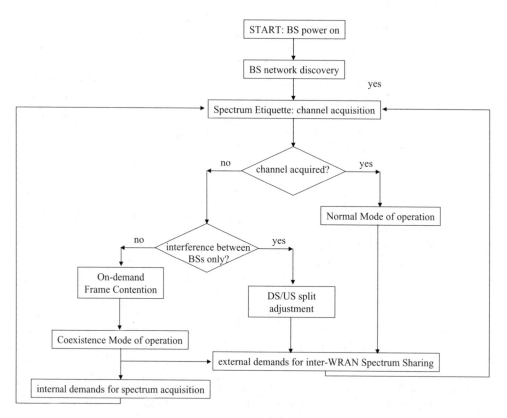

Figure 4.29 Execution flow of the inter-BS self-coexistence mechanism.

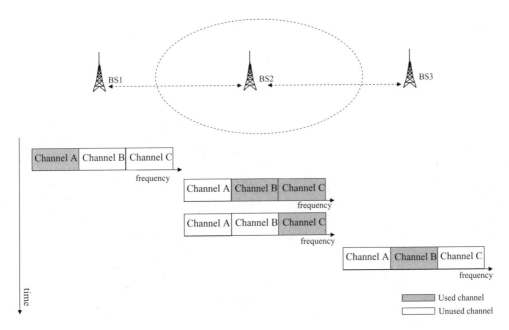

Figure 4.30   Example of dynamic resource sharing with interconnected WRAN base stations.

If the BS fails to acquire a free channel, it selects a channel occupied by one or more cells and identifies whether the interference source is another BS or the CPEs. In the first case, it runs a DS/UP split adjustment; otherwise it runs the on-demand frame contention algorithm and turns to the self-coexistence mode of operation.

Internal demands for resource acquisition or external demands for inter-WRAN spectrum sharing bring the flow diagram again to the spectrum etiquette step. Dynamic resource sharing algorithms enable the cooperation among IEEE 802.22 cells to optimize the usage of spectrum resources among coexistent WRAN cells. An example is the scenario shown in Figure 4.30, with interconnected base stations. Spectrum can be shared among cells of the same operator (the intra-operator situation) or cells of different operators (the inter-operator situation).

Figure 4.30 depicts the example where, at the beginning, BS1 is using channel A and is not using channels B and C. Because of that, BS1 offers the unused channels to BS2, which then transmits on channels B and C. After a while, BS2 does not need channel B any more, but it is requested by BS3 which can use it for transmission. In this example, orthogonal channels in frequency and time are used and all the BSs run in normal mode of operation.

Coexistence Beacon Protocol

The coexistence Beacon Protocol (CBP) is a protocol that allows self-coexistence among IEEE 802.22 cells. In self-coexistence mode, two or more IEEE 802.22 cells use the same channel. The coexistence beacon protocol (CBP) is a best-effort protocol based on the transmission of CBP packets carrying key information to adjacent and overlapping IEEE 802.22 cells for self-coexistence handling.

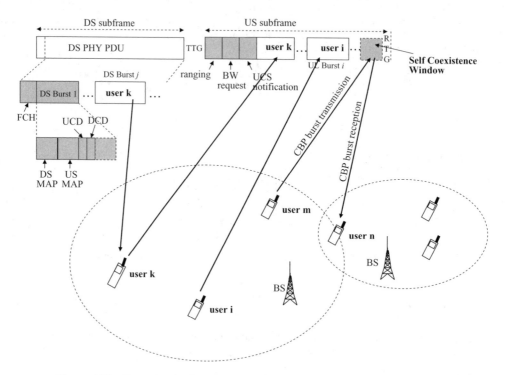

Figure 4.31 Example of a CBP burst transmission on the self-coexistence window.

A WRAN running in normal mode of operation turns to self-coexistence mode when it detects and decodes a superframe control header (SCH) or a coexistence beacon protocol (CBP) burst from an adjacent WRAN cell. CBP bursts are transmitted in the self-coexistence window (SCW) at the end of the uplink MAC frame. The source of a CBP packet can be a BS or a CPE; CBP packets can also be sent over the backhaul encapsulated in an IP packet. Figure 4.31 shows an example of CBP packet transmission and reception in the SCW.

The structure of a CPB packet, which is a CBP physical data unit (PDU), is represented in Figure 4.32. The CBP burst starts with a preamble and is followed by a CBP MAC PDU. The CBP MAC PDU includes a superframe control header (SCH) field containing the MAC address of the network element (CPE or BS) transmitting in the SCH. The SCH also advertises the schedule of quiet periods (QPs) and self-coexistence windows (SCWs) to CPEs in other neighbouring cells. The SCH is followed by the frame number and a header check sequence (HCS).

If the neighbouring cells operate on different TV channels, the reception of the SCH is used from other WRANs to discover quiet periods (QPs) for out-of-band sensing. If the neighbouring cells operate on the same TV channel or on adjacent channels, the SCH is used to schedule QPs and SCWs. The CBP burst can also contain other information elements (IEs) used for spectrum management, frame management, geo-location and CBP frame security under a coexistence situation.

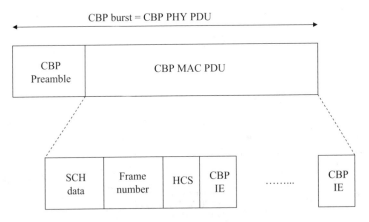

Figure 4.32   CBP packet structure.

Spectrum Etiquette

Spectrum etiquette occurs when there are enough channels available and allows selection of distinct operating and first backup channels. The information on operating, backup and candidate channels of each cell is exchanged through the use of CBP packets among the WRAN cells. The flow chart of the spectrum etiquette process for a channel selection decision is represented in Figure 4.33.

Spectrum etiquette is triggered by the following events:

- Incumbent discovery
- Neighbour WRAN cell discovery or update
- Operating channel switch demand (i.e., because of interference)
- Contention request received from neighbour WRAN cells

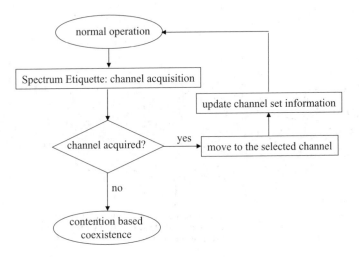

Figure 4.33   Flow chart of the spectrum etiquette process.

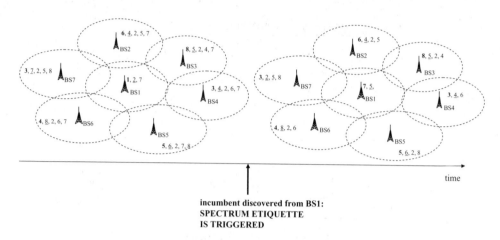

incumbent discovered from BS1:
**SPECTRUM ETIQUETTE
IS TRIGGERED**

Figure 4.34   Example of the spectrum etiquette procedure.

An illustration of an example of the spectrum etiquette procedure follows [12]. The central cell of BS1 has six neighbour cells (BSs 2, . . . , 7). In each cell, the operating channels are in bold, the backup channels are underlined and the others are candidate channels. For example, the central cell operates on channel 1, has channel 2 as backup and channel 7 is candidate. At a certain time, BS1 discovers an incumbent transmission on its operating channel and the spectrum etiquette is triggered to change both operating and backup channels. The candidate channel 7 becomes operating and channel 5 becomes a backup. BS1 sends channel set update information to the neighbour cells and the other BSs start the spectrum etiquette procedure. Figure 4.34 shows this spectrum etiquette procedure.

If another incumbent appears on channel 7 of BS1, channel 5 is promoted as the operating channel and BS1 starts contention-based self-coexistence with BS5, which operates on channel 5 as well. Figure 4.35 shows the case with contention-based self-coexistence triggering.

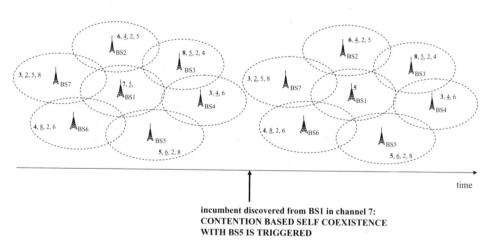

incumbent discovered from BS1 in channel 7:
**CONTENTION BASED SELF COEXISTENCE
WITH BS5 IS TRIGGERED**

Figure 4.35   Example of contention-based self-coexistence triggering.

On Demand Frame Contention (ODFC)

When contention-based self-coexistence is triggered, the on demand frame contention (ODFC) protocol is used to resolve contentions of frame resource among neighbouring base stations. A BS in self-coexistence mode will schedule one or more self-coexistence windows (SCWs) to monitor potential frame requests coming from WRAN cells using the same channel. The overlapping WRAN cells can also schedule other SCWs and can contend for frames used by the original WRAN cell by sending control messages through CBP bursts in these SCWs.

The WRANs in self-coexistence mode that operate on the same channel may each reserve its own SCWs or may contend the CBP burst transmission on the same SCWs. In the first case, the CBP burst transmissions are contention free; in the second case, the total overhead is reduced. However, the BSs can dynamically change the SCW scheduling to adapt to the self-coexistence exigencies.

Once the CBP bursts received from the BS or from its associated CPEs are decoded, the on demand frame contention (ODFC) protocol is applied to determine how the distribution of subsequent frames and superframes must be changed to adapt to the self-coexistence BSs exigencies. The control messages of the ODFC are carried from the CBP information elements. The message flow of the ODFC protocol is shown in Figure 4.36.

In Figure 4.36, the frame contention source (FC_SRC) sends a frame contention request (FC_REQ) message carrying the MAC address of the frame contention destination BS, a random number (frame contention number N_c) indicating the priority for the frame contention on that channel and the list of frames within a superframe that the contention source requests to acquire.

The frame contention destination (FC_DST) answers to the request with a response message (FD_RES) containing the source identifier and the list of granted frames. The FC_RES also carries the frame release time information, which encodes the number of superframes after which the channel will be released from the FC_DST. In the acknowledge (FC_ACK), the list of frames that the source has won through the frame contention algorithm, which it will use starting from the next superframe, is carried. Finally, the release message (FC_REL) notifies the list of frames that will be released to the winning FC_SRC. Figures 4.37 and 4.38 show two examples of the ODFC messages exchange in the two cases of contention source and contention destination success.

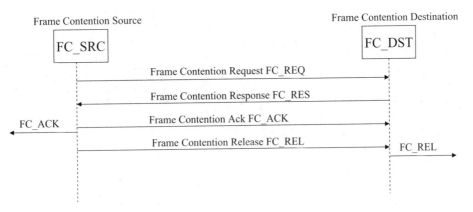

Figure 4.36   Message flow of the ODFC protocol.

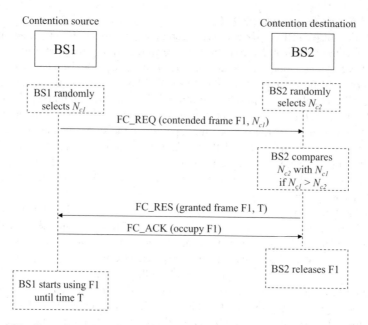

Figure 4.37 Example of ODFC message exchange in the case of contention source success.

In the example of Figure 4.37, BS1 is the contention source and BS2 is the contention destination. BS1 and BS2 randomly select the contention numbers N_{c1} and N_{c2}. BS1 transmits a contention request for frame F1. BS2, after having verified that N_{c1} is higher than N_{c2}, grants frame F1 for use during time T. BS1 acknowledges the occupation of frame F1 for the granted time.

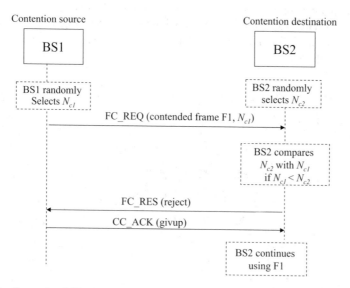

Figure 4.38 Example of ODFC message exchange in the case of contention destination success.

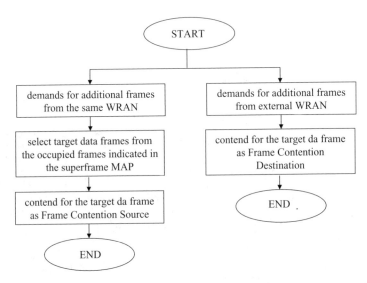

Figure 4.39   Top-level procedure of the on demand frame contention protocol.

In the example of Figure 4.38, BS1 is the contention source and BS2 is the contention destination. BS1 and BS2 randomly select the contention numbers N_{c1} and N_{c2}. BS1 transmits a contention request for frame F1. BS2, after having verified that N_{c1} is lower than N_{c2}, rejects frame F1 for use during time T. BS1 acknowledges giving up frame F1 and BS2 continues using F1.

The top-level procedure of the on demand frame contention protocol is shown in Figure 4.39. Each BS executes the ODFC procedure if it receives an internal or external request for new frames.

4.3   IEEE 802.22.1

In 2010, IEEE published the 802.22.1 standard with the goal to enhance harmful interference protection for low-power licensed devices operating in TV broadcast bands [19].

The spectrum holes left in the TV bands, called 'white spaces', can be used for the deployment of new unlicensed wireless services. The unlicensed devices operate on the unused portions of the TV spectrum without interfering with the incumbent's transmissions. Among these transmissions are included low-power licensed devices like wireless microphones operated by broadcasters.

The IEEE 802.22.1 standard describes the air interface of beacon devices, whose aim is to protect low-power licensed devices like microphones. The beacons contain synchronization bursts and, as optional additional information, locations and operational parameters of the protected devices. The beacons should be installed by the operators of the licensed devices; unlicensed systems should include an IEEE 802.22.1 receiver able to decode the beacons and use such information in order to assure interference protection to the licensed systems.

The principal element of the beaconing network is the protecting device (PD), which can be a primary protecting device (PPD) or a secondary protecting device (SPD). The PPD is a

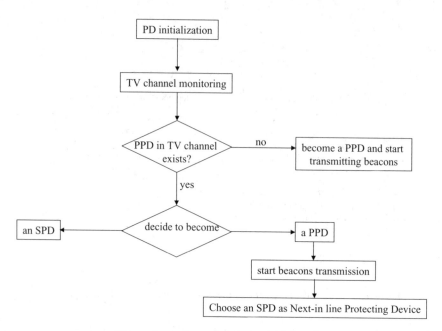

Figure 4.40   Protecting device initialization.

device that periodically transmits beacons to protect its corresponding licensed devices. The SPD uses beacons to transmit with a PPD and helps the PPD in licensed devices protection. The next-in-line protecting device (NPD) is an SPD that becomes a PPD if the existing PPD stops beaconing.

During the initialization, a PD monitors its TV channel to determine if there is a PPD transmitting beacon. If a PPD is not found, then the PD acts as a PPD and periodically transmits beacons. If one or more PPDs are found, the PD can act as a PPD and start beacon transmission or it can act as an SPD and contact a PPD. PPD chooses an SPD to become an NPD. The described initialization procedure is shown in Figure 4.40.

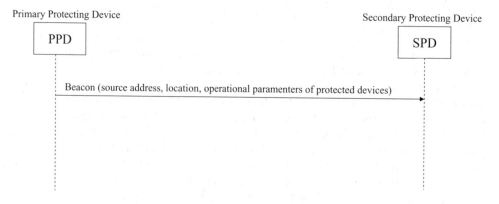

Figure 4.41   PPD beaconing.

Primary Protecting Device   Secondary Protecting Device

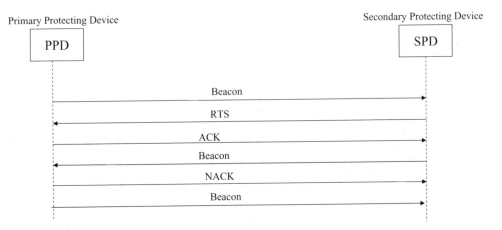

Figure 4.42 SPD communication with PPD.

The PPD transmits the beacons in broadcast using an aloha access mechanism with information about the licensed device, like the location and estimated duration of TV channel occupancy. The beacon reception is not acknowledged from the SPDs.

The SPDs can also transmit information to the PPDs. In this case, the communication starts with a request to send (RTS) message. If the PPD receives the RTS and sends an ACK, it suspends the beacon transmission and the SPD transmits a beacon. If the PPD answers the SPD beacon with a NACK, the PPD resumes its normal transmission of beacon frames. PPD and SPD communications are shown in Figures 4.41 and 4.42.

References

1. Yücek, T. and Arslan, H. (2009) A survey of spectrum sensing algorithms for cognitive radio applications. *IEEE Communications Surveys and Tutorials*, **11** (1), 116–130.
2. Spectrum Sensing, End to End Efficiency (E3) White Paper, November 2009.
3. Chouinard, G. (2011) The IEEE 802.22 WRAN Standard and Its Interface to the White Space Database, IETF PAWS Working Group Meeting, Québec City, 26 July 2011.
4. Federal Communications Commission (2008) In the Matter of Unlicensed Operation in the TV Broadcast Bands: Second Report and Order and Memorandum Opinion and Order. Technical Report 08-260, November 2008.
5. Mlinarsky, F. (2011) The Role of White Spaces in the Realm of Wireless Broadband, 8 June 2011, Octoscope Inc.
6. Federal Communications Commission (2010) In the Matter of Unlicensed Operation in the TV Broadcast Bands: Second Memorandum Opinion and Order. 23 September 2010.
7. Federal Communications Commission (2012) In the Matter of Unlicensed Operation in the TV Broadcast Bands Additional Spectrum for Unlicensed Devices Below 900 MHz and in the 3 GHz Band: Third Memorandum Opinion and Order, 5 April 2012.
8. ECC (2011) Technical and Operational Requirements for the Possible Operation of Cognitive Radio Systems in the 'White Spaces' of Frequency Band 470–790 MHz, Report 159, January 2011.
9. ECC (2013) Futher Definition of Technical and Operational Requirements for the Operation of White Space Devices in the Band 470–790 MHz, Report 185, Complementary Report to ECC Report 159, January 2013.
10. ECC (2013) Technical and Operational Requirements for the Operation of White Space Devices under Geo-location Approach, Report 186, January 2013.

11. Meeting Document, Combination of Geo-location Database and Spectrum Sensing Techniques. 9th Meeting of SE43, Copenhagen, 14–15 April 2011.
12. IEEE P802.22-2011 Standard (2011) *Wireless Regional Area Networks*, Part 22: Cognitive Wireless RAN Medium Access Control (MAC) and Physical Layer (PHY) Specifications: Policies and Procedures for Operation in the TV Bands, July 2011.
13. 22-10-0073-03-0000, IEEE 802.22 Wireless Regional Area Networks.
14. Stevenson, C.R., Chouinard, G., Lei, Z., Hu, W., Shellhammer, S.J. and Calwell, W. (2009) IEEE 802.22: The First Cognitive Radio Wireless Regional Area Network Standard. *IEEE Communication Magazine*, January 2009, 130–138.
15. Rose, M. and McCloghrie, K. (1990) Structure and Identification of Management Information for TCP/IP-based Internets. *RFC* **1155**, May 1990.
16. Rose, M. and McCloghrie, K. (1991) Management Information Base for Network Management of TCP/IP-based Internets: MIB-I. *RFC* **1213**, March 1991.
17. Case, J., Fedor, M., Schoffstall, M. and Davin, J. (1990) A Simple Network Management Protocol (SNMP). RFC 1157, May 1990.
18. 22-10-0121-02-0000, IEEE 802.22 Coexistence Aspects.
19. IEEE Standard for Information Technology (2010) *Telecommunications and Information Exchange between Systems, Local and Metropolitan Area Networks – Specific Requirements*, Part 22.1: Standard to Enhance Harmful Interference Protection for Low-Power Licensed Devices Operating in TV Broadcast Bands, 1 November 2010.

5

ETSI Standards on Reconfigurable Radio Systems

5.1 Introduction

The access network evolution has led in recent years to scenarios in which a single IP-based backbone provides the transport service to multiple access networks, both fixed and wireless. Operators have to manage many radio access technologies (RATs), often by 'manually configuring' radio parameters. Recently, some common radio resource management (CRRM) features have been introduced in the multi-RAT network in order to optimize network utilization.

Common radio resource management (CRRM) deals with a unique radio resource management (RRM) that controls different radio access technologies, for example GSM at 900 MHz and 1800 MHz, UMTS at 2.1 GHz and LTE, with different cell layers (macro, micro, pico, etc.). In general, a CRRM tool chooses the optimum layer depending on terminal capability, user profile and cell load, in order to optimize network performances.

However, CRRM manages a multi-RAT access network through a static configuration of parameters. The evolution of CRRM adds more intelligence in the access network, with self-organizing networks (SONs), software defined radio (SDR) and cognitive radio (CR). The development of cognitive radio terminals and networks will introduce new scenarios in the radio access network.

All over the world a refarming process of the radiomobile spectrum is going on. In the near future, radiomobile frequencies will not be strictly associated to a technology, but they will be used depending on user terminals, service profiles, traffic requests and network optimization.

With spectrum sensing cognitive radio terminals and networks, spectrum usage will be optimized among different radio access technologies. Full cognitive radio, in which every possible parameter observable by a wireless node or network is taken into account for adaptation and network optimization, is evolving towards reconfigurable radio systems (RRSs), on both the network and terminal sides [1, 2].

Reconfigurable Radio Systems: Network Architectures and Standards, First Edition. Maria Stella Iacobucci.
© 2013 John Wiley & Sons, Ltd. Published 2013 by John Wiley & Sons, Ltd.

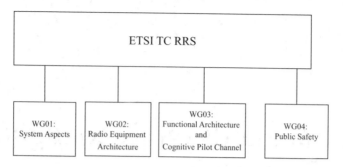

Figure 5.1   ETSI Technical Committee structure.

The standardization of reconfigurable radio systems (RRSs) is done in ETSI's RRS Technical Committee (TC), with the following four working groups (WGs):

- RRS 01 System aspects
- RRS 02 Radio Equipment Architecture
- RRS 03 Functional Architecture and Cognitive Pilot Channel
- RRS 04 Public Safety

Figure 5.1 shows the ETSI Technical Committee structure.

This chapter describes the ETSI proposals for reconfigurable radio systems (RRSs).

5.2   ETSI Reconfigurable Radio Systems

In 2009, ETSI started to publish technical reports on RRS, including a functional architecture [3] for RRSs. The ETSI RRS functional architecture proposes single-operator and multioperator scenarios, with new entities for dynamic spectrum management, dynamic self-organizing network planning and management, joint radio resource management and configuration control (CC). The new entities bidirectionally communicate through standard interfaces and with their corresponding system in the radio terminal. Moreover, the ETSI RRS functional architecture proposes the introduction of a cognitive pilot channel (CPC) for the management of the radio context (i.e., frequencies, operating RATs, etc.) [4]. The considered scenario is with heterogeneous access networks connected with a unique packet-based core network, as represented in Figure 5.2.

Radio access networks (RANs) operate with various technologies at different frequencies, with multiple coverages of macro, micro and pico cells. They work with many types of terminals, base stations (BSs) and access points (APs). Terminals can be a legacy, multistandard, cognitive radio; base stations can be a legacy, cognitive radio, reconfigurable; access points are in general legacy APs. Terminals can communicate through a direct radio link (peer devices) or through a connection to a BS, can support multiple connections and can operate as relays to other terminals. Base stations realize the radio coverage, transmit with the terminals and operate if necessary as wireless relays for other BSs.

Figure 5.3 shows an example of a heterogeneous wireless network scenario.

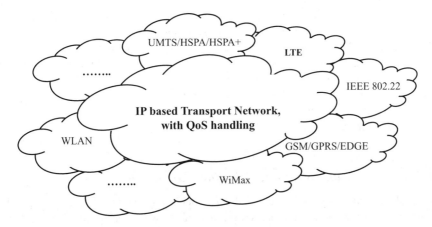

Figure 5.2   Heterogeneous access networks scenario.

In the considered scenario the radio access networks can belong to the same operator or to different operators and can be associated with the same transport network or different transport networks [3]. Some of the radio access network nodes are able to reconfigure their radio parameters, whose management is handled from the functional architecture (FA) described in this chapter. Radio frequencies can also be dynamically managed, with specified rules for spectrum usage and allowed maximum powers. The FA even enables spectrum sharing, spectrum renting and primary and secondary spectrum usage.

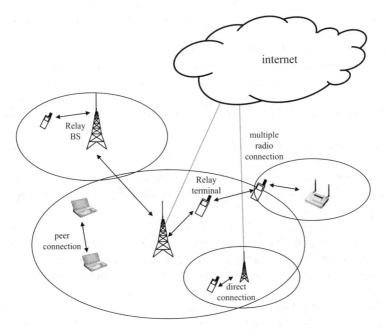

Figure 5.3   Example of a heterogeneous wireless network scenario.

The FA sends to the terminals the management information related to which radio accesses are available at a given location and time, in order to optimize both the usage of the network and the spectrum. The FA receives from the terminals management information related to the detected RANs, terminal capabilities, etc. The FA also bidirectionally communicates with the underlying RANs in order to obtain information about the BSs capabilities and to manage the radio parameters of reconfigurable BSs. Reconfigurable radio base station (RBS) implementation aspects are described in [5], and include basic and reconfigurable RBS architectures, summarized in the next section.

5.2.1 Reconfigurable Radio Base Station Architecture

The availability of reconfigurable nodes in the network will give to the operators the possibility to manage the radio resources in order to optimize the network and spectrum utilization. In a mature network deployment scenario, the RBSs will also be provided with self-establishment and self-optimization functionalities.

A reconfigurable RBS has the following basic requirements [5]:

- Multistandard: the RBS implements more than one standard radio access technology (RAT), that is GSM/GPRS/EDGE, UMTS/HSPA/HSPA+ and LTE/LTE advanced. The different RATs are implemented as software applications.
- Self-establishment capabilities: when switched on, the RBS determines the initial radio configuration. It sets the output power, the electrical tilt for coverage, configures physical resources and maps logical channels, generates the cell identifier (cell-Id), downloads the neighbour list from an external database and sets up radio parameters for congestion and admission control, handover thresholds, etc. It also establishes the connection with the core network elements.
- Spectrum flexibility: the RBS is able to work with different RATs at different frequencies, which can be dynamically changed. For example, because of spectrum refarming, some LTE carriers could be moved from 2600 to 1800 MHz and some UMTS carriers from 2100 to 900 MHz.
- Spectrum trading: the RBS is capable, if allowed from spectrum management rules, to change the use of the spectrum while maintaining the right to use. For example, secondary markets could be enabled through the possibility of giving permission to secondary users to use the same RBSs frequencies.
- Self-optimization capabilities: the RBS continuously performs coverage and capacity optimization, neighbour list updating, handover optimization, interference control, energy saving, etc.

The reconfigurable radio base stations will be able to communicate through standard interfaces with the network management system in order to execute the received reconfiguration commands quickly and to send reports about traffic measurements, network configuration, etc. The reconfigurable RBS architecture evolves from the basic RBS architecture, built of functional blocks, and represented in Figure 5.4 [5,6].

A block represents a logical grouping of a set of functions and can be spit into one or more modules, each implementing a subset of functions. The functional blocks communicate through

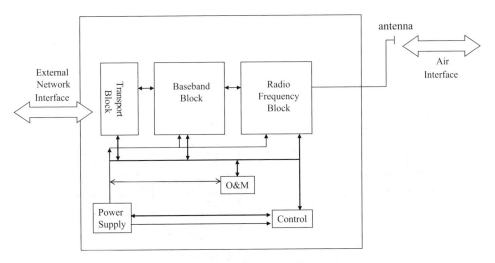

Figure 5.4   Basic RBS architecture.

internal interfaces; some of them transport data and signalling through external interfaces, like the network or the air interface.

The external network interface allows the communication of the RBS with other access network nodes, like the base station controller (BSC) in GSM/GPRS/EDGE with the Abis interface; radio network controller (RNC) in UMTS/HSPA/HSPA+ with the IuB interface; and evolved node Bs (eNBs) in LTE with the X2 interface. The air interface allows the communication of the RBS with the user equipment (UE) with standard radio interfaces (i.e., UMTS-Uu, LTE-Uu). Internal interfaces exchange control, status and alarms between the control and clock block and other RBS blocks.

The basic RBS architecture blocks are [5, 6]:

- Transport block (TB): realizes the connection with the access network nodes. The TB conveys user plane, control plane and management plane information among the network interface and the other RBS functional blocks.
- Control block (CB): is the block that manages the configuration and the radio resources of the RBS.
- O&M block: performs RBS operation and maintenance and interacts with the network management system (NMS) or element management system (EMS).
- Baseband block (BB): performs baseband processing of the transmission chain, like channel encoding, interleaving, rate matching, MIMO precoding, etc. It can be realized with more than one module, each performing baseband processing of a standard transmission chain (i.e., GSM, UMTS, LTE, WiMAX, etc.).
- Radio frequency block (RFB): is the block connected to the antenna that handles the radio frequency functions, like up/down conversion, carrier selection, power amplification, transmission and reception diversity, etc.

The reconfigurable RBS architecture evolves from the basic RBS architecture. In a reconfigurable RBS, the control block is added to a number of features related to the joint

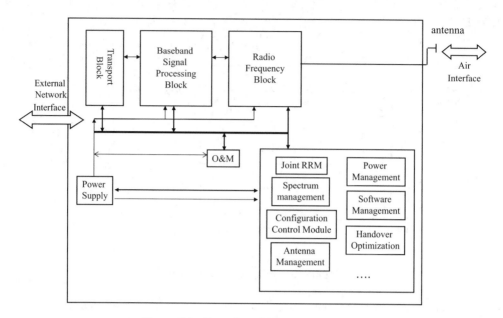

Figure 5.5 Reconfigurable RBS architecture.

management of radio resources, like spectrum, antenna, power and software (SW), and to the control of the RBS multi-RAT, with global optimization purposes. The reconfigurable RBS architecture is represented in Figure 5.5.

In the figure, the control module of the reconfigurable RBS includes the following blocks:

- Power management: defines the maximum allowed powers per radio access technology, per cell, per sector, per antenna. Other than fixing the power limits, the power management block implements energy efficiency. With energy efficiency, it is possible to adaptively switch-off or put in stand-by mode the resources that are not used because of the actual amount of traffic, thus saving energy and money.
- Software management: this block is responsible for the RBS software download and installation, software activation/deactivation, etc.
- Spectrum management: it regulates the use of radio frequencies to optimize the usage of the assigned frequencies, for primary and secondary usages.
- Configuration control module: controls the changes of configuration parameters of each radio access technology. It also manages and solves faults through reconfiguration actions.
- Handover optimization: this function has the objective of minimizing handover failures, minimizing unnecessary handovers and increasing the load balancing capability. It concerns intrafrequency, interfrequency and inter-RAT handover.
- Antenna management: this bock controls the antenna parameters, like gain, radiation pattern, polarization, MIMO configuration, etc.
- Joint radio resource management (JRRM): this block manages and controls the radio resources of each radio access technology through a joint optimization of admission control, scheduling and other control (i.e., cell load, cell capabilities, etc.) tasks.

Some functions of the reconfigurable RBS can be combined with others. For example, software management, power management and handover optimization can be included in the JRRM entity.

5.2.1.1   Use Cases for a Reconfigurable Radio Base Station

The scenario considered for the proposed use cases is a heterogeneous radio access network managed from the same operator. The RBSs in this scenario support different RATs implemented as software applications, which can also be installed and uninstalled during operation. Spectrum sharing functionalities allow spectrum resources to be distributed dynamically among the radio access technologies with the goal of achieving radio resources optimization.

The entities involved in the given scenario are: terminal equipments (TEs), radio base stations (RBSs), joint radio resource management (JRRM) entity and configuration control module (CCM). The RBSs are reconfigurable nodes equipped with hardware processing resources (HPRs) shared among different RATs. Each RAT is a radio application using a certain amount of HPRs depending on the configuration of radio parameters, the amount of on-going traffic, etc. The HPR is connected to the radio frequency (RF) block, which goes to the antenna and is able to support different RATs at the same time. The JRRM entity jointly handles the RAT radio resources, performing radio bearer control, radio admission control, scheduling of resources in uplink and downlink, load balancing, etc. The configuration control module has the task of evaluating possible RBS reconfigurations. In the described use case, RBSs, JRRM and CCM are three distinct entities.

The first proposed scenario is an intrasystem radio resources reconfiguration, with reconfigurable RBSs belonging to the same operator and where the spectrum assigned to RAT1 changes according to traffic needs (spectrum sharing). Figure 5.6 shows the high level information flow related to the case of the intrasystem radio resources reconfiguration.

In the figure, RAT1 (i.e., UMTS) of RBS1 is using band B1 (i.e., two carriers of 5 MHz); RAT1 (i.e., UMTS) of RBS2 is using band B2 (i.e., one carrier of 5 MHz). The radio resources optimization function of the JRRM entity decides that spectrum sharing between the two RBSs is needed and the reconfiguration algorithm of CCM determines that one carrier must be added to RBS2 and one carrier must be removed from RBS1. RBS reconfiguration commands are sent to RBS1 and RBS2 and the spectrum is successfully reassigned between the two base stations. If the network is provided with a cognitive pilot channel (CPC), it is updated with the new context information (i.e., spectrum rearrangement between RBS1 and RBS2). Figure 5.7 shows the use case of flexible spectrum allocation between different RATs of the same RBS.

In the figure, the reconfigurable radio base station (RBS) is running RAT1 (i.e., UMTS) in bandwidth B1 and RAT2 in bandwidth B2 (i.e., GSM) together. The radio resources optimization function of the RRM entity decides to reallocate the spectrum between RAT1 and RAT2. The CCM reconfiguration algorithm decides that one or more carriers must be added to RAT2 and one carrier must be removed from RAT1. An RBS reconfiguration command is sent to RBS1 and the spectrum is successfully reassigned between the two radio access technologies. If the network is provided with a cognitive pilot channel (CPC), it is updated with the new context information.

The third considered use case is the RBS radio software upgrade, for example to upgrade an existing RAT (i.e., add HSPA+ to HSPA) or to install a new RAT. The software will

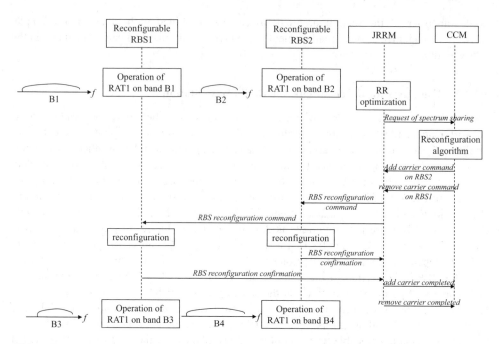

Figure 5.6   High level information flow for the intrasystem radio resources reconfiguration use case.

be downloaded from a software database, held by an entity that can be the operator, the manufacturer or a radio application provider. The software database could also be in the configuration control module (CCM).

In the example shown in Figure 5.8, the new functionality deployment decision comes from the configuration control module, up to the RBS through the JRRM entity. Then the RBS asks for new software downloading to the software database. When the software download is completed, the CCM commands the RBS to activate the new software functionality. When the new software is activated, the CCM runs an optimization algorithm, which implies an RBS reconfiguration. The RBS reconfiguration will be different if the new software is related to the upgrade of an existing RAT or if a new RAT is operating in the RBS. If the network is provided with a cognitive pilot channel (CPC), it is updated with the new context information (i.e., new functionality or new RAT in operation, eventually with updates to the used frequencies).

5.2.2   Reconfigurable Radio Device Architecture

The availability of reconfigurable multistandard RBSs in the network brings the need for multistandard mobile devices that are able to follow the network parameters reconfiguration and the changes in spectrum utilization.

One method to implement a multistandard mobile device is to design multiple radio frequency (RF) chains connected to multiple baseband circuits, as represented in Figure 5.9. For each radio access technology, there is a baseband module connected with a radio frequency

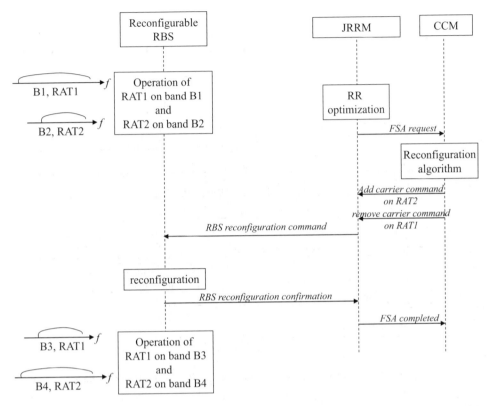

Figure 5.7 High level information flow for flexible spectrum allocation between different RATs of the same RBS.

block (RFB). The baseband module digitally processes baseband operations; the RFB performs up/down conversion, modulation/demodulation and filtering, and is connected to the antenna or the antennas. In this case it is not possible to update or enhance the device and the only chance is to activate the preconfigured transmission chains.

An alternative and more efficient solution to implement a multistandard mobile device is to use a software defined radio (SDR). The SDR device is adapted, updated and enhanced by using software, thus enabling reconfigurable radio system architectures. An SDR device implements more than one standard radio access technology (RAT) through different software radio applications that can be downloaded, installed and activated. Moreover, the device is able to simultaneously execute multiple radio interfaces operating at different frequencies through the execution of different radio software applications.

The scheme of Figure 5.9 evolves towards the scheme of Figure 5.10, with a baseband module connected to a soft transceiver implementing digital up/down conversion and modulation/demodulation. The soft transceiver is then linked to the multiband radio front end module (MBFEM) with power amplifiers and tunable filters. The MBFEM is connected to the antenna.

ETSI proposed in [7] an SDR reference architecture for a mobile device, where the device is a radio computer with a radio operating system, a physical radio computing platform

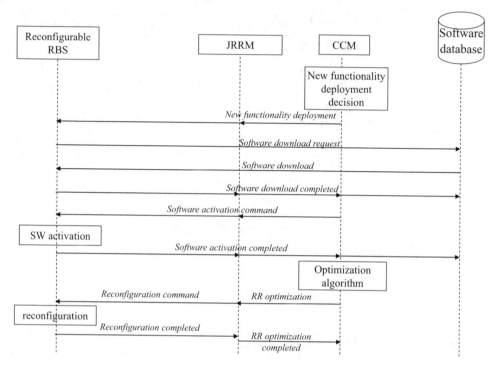

Figure 5.8   High level information flow for an RBS radio software upgrade.

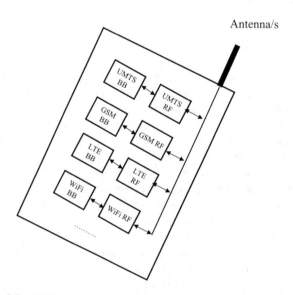

Figure 5.9   Multistandard mobile device with multiple transmission chains.

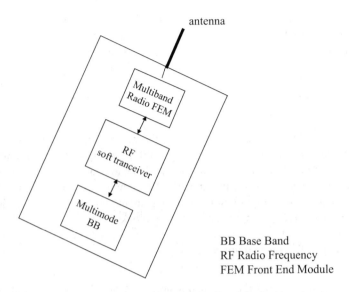

BB Base Band
RF Radio Frequency
FEM Front End Module

Figure 5.10   SDR multistandard mobile device.

and radio applications realizing multiple RATs. The radio computer concept is shown in Figure 5.11.

The radio applications can be installed, updated and enhanced like any other software in a computer. The multitask radio operating system runs multiple radios working at the same or different frequencies. Other than the radio operating system and radio applications, the radio computer is provided with a physical platform realizing the following stages:

- Antennas, which are the radiating elements. The radio computer can be provided with many narrowband antennas operating at different frequency bands or with a wideband antenna with tuners. Figure 5.11 shows a broadband antenna with tuning.

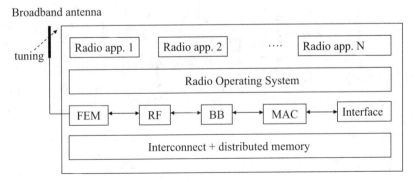

Figure 5.11   Radio computer concept.

- Front end module (FEM): includes filters, mixers and power amplifiers.
- Radio frequency (RF) module, containing a digital RF transceiver.
- Baseband module: it digitally processes all the baseband functions of the transmission/reception chain of the different active radio access technologies.
- MAC: performs medium access control.
- Interface: interfaces the physical platform to the radio applications.

Multiple radios can be run simultaneously in the radio computer, sharing the operating system and the physical platform, but with the possibility to adjust, for each running technology, the radio parameters like output power, modulation and coding scheme, and the operating frequency channels. The proposed functional SDR architecture for a mobile device [7, 8] includes functional blocks for configuration, management and control, unified radio applications and standard interfaces between the functional blocks. The functional architecture for an SDR device is represented in Figure 5.12.

The functional blocks in the architecture are [9]:

- Administrator: is responsible for radio applications installation and disinstallation; creates and deletes instances of radio applications; manages the configuration of radio application instances.

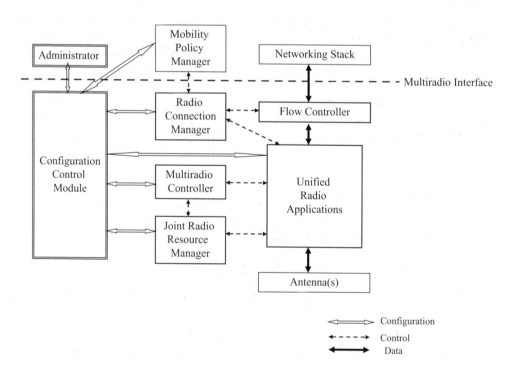

Figure 5.12   Functional architecture for an SDR device.

- Mobility policy manager: manages the activation and deactivation of radio application instances to allow the mobile terminal to move from one radio access technology to another; manages the peer devices discovery, association and disassociation, and peer connections.
- Networking stack: implements the layers of the protocol stack for the transmission and reception of data flow.

The interface between the administrator, mobility policy manager and networking stack is the multiradio interface.

The radio computer architecture includes the following functional blocks:

- Configuration control module (CCM): is responsible for the radio computer configuration: installation, disinstallation and loading and unloading of radio applications. It configures and manages, for each application, the radio parameters.
- Radio connection manager (RCM): it manages the radio connection of user data through the activation or deactivation of data transmission on one or multiple RATs.
- Flow controller (FC): controls the user plane data.
- Multiradio controller (MRC): it controls the requests for spectrum resources arriving from different working radio applications in order to prevent interferences and conflicts.
- Joint radio resource manager (JRRM): manages and optimizes the usage of radio computer resources, shared among many active applications.

The radio applications communicate with all other radio computer functional blocks through a unified radio application interface.

Another two interfaces are needed for the radio computer operation:

- The radio programming interface, between the radio software entities and the radio computer platform
- The interface to the RF digital transceiver

The described functional architecture of the reconfigurable mobile device is basically a software defined radio (SDR) architecture [7], with software modules handling the multi-RAT transmission and reception. The SDR architecture of the mobile device can be added to cognitive functionalities [9] (CF), built on top of the SDR subsystem and leading to a reconfigurable radio device (RRD). The reconfigurable radio device functional architecture is represented in Figure 5.13.

The separation of SDR and cognitive functionalities allows independent development of the relative subsystems. The cognitive functionalities (CFs) will implement the intelligence necessary to discover the presence of radio networks and peer mobile terminals, to receive context information, to select an available radio access technology, etc.

The functional architecture for the reconfigurable radio device, made of a radio operating system, a physical radio computing platform, radio applications and the cognitive module implies that new actors are come on to the market scenario. Moreover, because the radio applications can be downloaded over the air, new entities responsible of the over the air (OTA) radio application provisioning will come on to the new market scenario.

The possible actors in the new market scenario are shown in Figure 5.14. In the figure, the customer is in the centre of the graph. It uses a reconfigurable radio device communicating

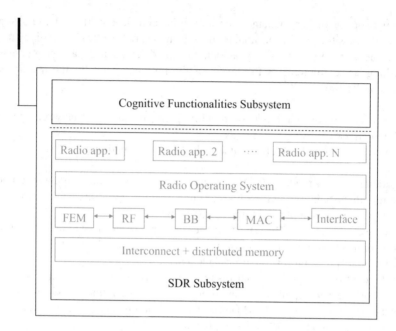

Figure 5.13   Functional architecture for the reconfigurable radio device.

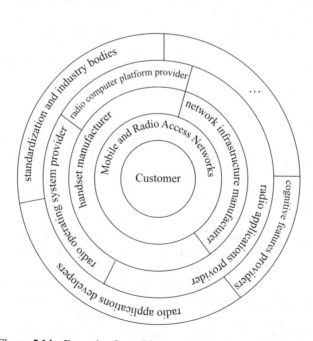

Figure 5.14   Example of possible actors in the new market scenario.

with mobile and radio access networks, handled from one or more network operators. To enable this scenario, probable old and new actors are: network operators, handset manufacturers, network infrastructure manufacturers, radio computer platform providers, radio operating system providers, radio application providers, radio application developers, cognitive features providers and others. Standardization bodies will guarantee the compatibility and interoperability among all parts of the reconfigurable radio device, and between the reconfigurable device and the radio access networks; industry bodies will support the standardization, the deployment and the promotion of reconfigurable radio terminals and networks [10, 11].

A reconfigurable radio device (RRD), when powered on, switches on its radio access technologies by activating the corresponding radio applications. The RRD uses the radio interfaces to acquire context information, through the cognitive pilot channel or other techniques. The radio applications that are not needed are deactivated. The next step is the selection of one or more available RATs to establish one or more associations with other peer terminals or radio access network nodes for the exchange of data. Selection of the RATs can be performed from the mobile device or can be ordered from the network; then the radio parameters of the applications are properly configured. In the reconfigurable radio network scenario it is also possible to move a data flow from one association to another, even across different radio applications. The described logical flow for network entry, association and data flow transmission is represented in Figure 5.15.

Figure 5.15 also includes the sensing process, with the possibility of sensing requests coming from the network and the transmission of sensing results from the mobile terminal.

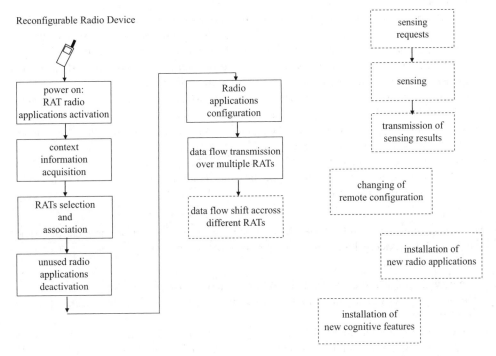

Figure 5.15   Logical flow for network entry, association and data flow transmission.

Other than sensing, the mobile device supports the installation of new radio applications and of new cognitive features on top of the SDR subsystem.

The logical flow represented in Figure 5.15 does not show the actions, defined from the standard of the involved radio accesses, related to the network entry process of the selected RATs. In this case, the logical flow includes actions such as scanning for downlink channels, synchronization with the serving base station, acquisition of uplink parameters, basic capabilities negotiation, authentication, registration to the network and IP address assignment.

5.2.2.1 Use Cases for the Reconfigurable Radio Device

ETSI proposed four use cases [12] for the RRD. The considered scenario is of heterogeneous radio access networks, belonging to the same or different operators, with reconfigurable radio capabilities.

The proposed use cases are related to:

- Terminal centric configuration: the RRD is able to detect the presence of the different radio access networks and to select, depending on the context, one or more of them for association and data transmission, with the goal of optimizing its operating conditions (i.e., battery consumption, memory, processing resources, etc.). This use case assumes that the different radio access technologies are not centrally managed. Because the optimization is not performed from the network, in this case the terminal makes choices to optimize its own resources.
- Network driven terminal configuration: this use case assumes that the heterogeneous access network is aware of the context of the terminal and its capabilities. In this case, the network decides the radio access technologies the terminal will be associated with and will use for data transmission. The network makes choices to optimize network performances and commands to the mobile the optimal configuration to be adopted.
- Addition of new features, such as support for novel radio systems, to RRDs: if the radio computer has not installed radio applications performing some new radio access technologies, new radio applications enabling the novel radio systems can be installed in the multiradio computer. Such new radio applications could also be downloaded from the network.
- Provision of new cognitive features: because the RRD needs to acquire the context related to the novel radio systems and their spectrum, new cognitive functionalities can be added on top of the SDR subsystem.

The scenario considered in [12] for the described use cases contemplate three entities, other than the mobile device, which is a reconfigurable radio device (RRD):

- The network infrastructure, operated from one or more network operators.
- The control point (CP): requests and receives information from the RRDs, like spectrum measurements, etc.
- The cognitive pilot channel (CPC) provision: is a channel that furnishes the context to the RRD, that is indications on the available RATs, frequencies used from the RATs, available frequencies for primary and secondary usage, etc.

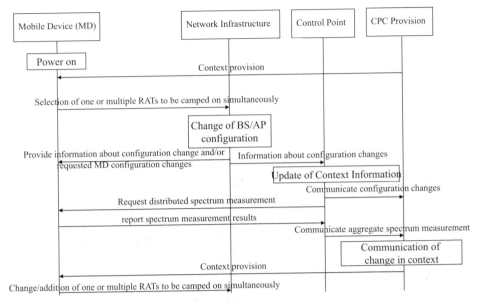

Figure 5.16 Information flow illustration for the use case of a terminal centric configuration in a heterogeneous radio context. (Reproduced with permission. Copyright European Telecommunications Standards Institute 2010. Further use, modification, copy and/or distribution are strictly prohibited. ETSI standards are available from http://pda.etsi.org/pda/.)

The control point and the CPC provision in [12] are separated entities with the goal of maintaining the maximum possible generalization. In real implementations, the control point and the CPC provision can be merged into one entity and managed from the network operator or from an external subject. The control point updates the context of the CPC provision with the information received from the terminals. Figure 5.16 illustrates the information flow for the first use case: the terminal centric configuration in a heterogeneous radio context.

The use case of a terminal centric configuration in a heterogeneous radio context requires that the mobile device (MD), after power on, acquires context information from a cognitive pilot channel (CPC) and/or from other mobile devices. Then the MD selects one or multiple RATs to be camped on (in other words, to identify the cell base station/radio coverage where the terminal is). The network infrastructure changes, if needed, the configuration of the involved base stations (BSs) and access points (APs), and communicates the changes to the control point (CP). The CP updates context information and communicates the changes to the CPC provision. The CP can also request the mobile device spectrum measurements in order to obtain the updated situation on frequency occupancy. The aggregated results are transmitted from the CP to the CPC provision, which updates the context and the information provided. If the MD decides to change or add new RATs to be camped on simultaneously, it communicates the decision to the network infrastructure. Figure 5.17 illustrates the information flow for the second use case: a network driven configuration in a heterogeneous radio context.

In this case, the selection of possible RATs and relative radio parameter configuration is driven from the network, and control point and CPC provision are optional entities, represented

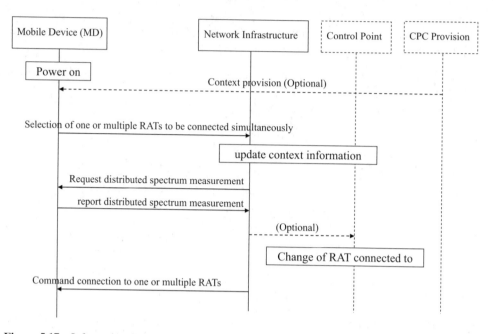

Figure 5.17 Information flow for a network driven configuration use case in a heterogeneous radio context. (Reproduced with permission. Copyright European Telecommunications Standards Institute 2010. Further use, modification, copy and/or distribution are strictly prohibited. ETSI standards are available from http://pda.etsi.org/pda/.)

in Figure 5.17 by dotted lines. The network selects suitable RATs for the mobile device, commands the MD to connect the selected RATs with a chosen configuration, and then updates the context. The network/CP can also request the mobile devices spectrum measurements in order to obtain the updated situation on frequency occupancy. The aggregated results are used to decide whether the MD must change the RATs to be connected in order to optimize network utilization; if so decided a new command for a RAT change and configuration is sent to the MD.

Figures 5.18 and 5.19 are related to the third and fourth use cases, regarding mobile device software updates. In the third use case the MD is updated with new features installed in the SDR subsystem, such as support for novel radio applications; in the fourth use case the MD is updated with new cognitive features (i.e., spectrum measurement of the novel radio system), installed on top of the SDR subsystem. The examples of Figures 5.18 and 5.19 show that the RRD detects, from the CPC, the opportunity to install new radio software.

5.2.3 Cognitive Pilot Channel (CPC)

The cognitive pilot channel (CPC) provides context information (i.e., operating radio access networks, radio access networks capabilities, geo-location information, spectrum occupancy, network measurements, mobile measurements, etc.) to support and facilitate the connectivity in the context of a heterogeneous radio access network. The CPC is then an enabler for different radio system coexistence and for dynamic spectrum access.

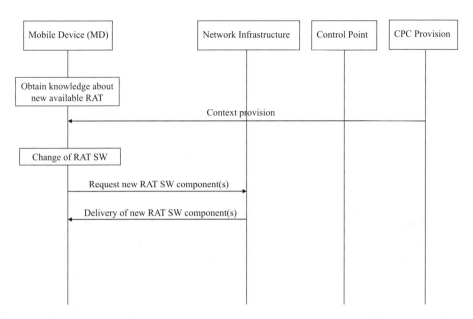

Figure 5.18 Example of provision of new features (i.e., support for novel radio systems). (Reproduced with permission. Copyright European Telecommunications Standards Institute 2010. Further use, modification, copy and/or distribution are strictly prohibited. ETSI standards are available from http://pda.etsi.org/pda/.)

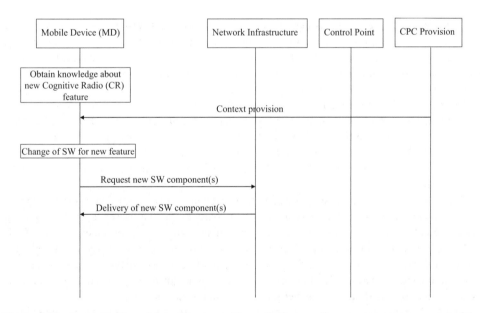

Figure 5.19 Example of provision of new cognitive radio features (i.e., spectrum measurement of the novel radio systems). (Reproduced with permission. Copyright European Telecommunications Standards Institute 2010. Further use, modification, copy and/or distribution are strictly prohibited. ETSI standards are available from http://pda.etsi.org/pda/.)

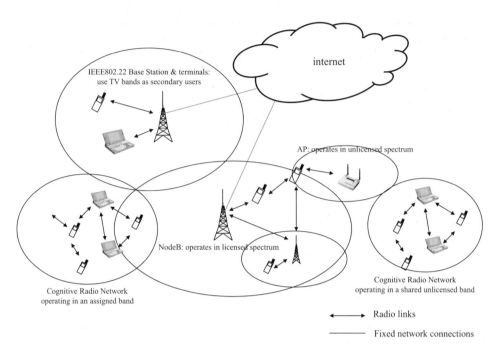

Figure 5.20   Heterogeneous network scenario with different spectrum usages.

The context of heterogeneous networks implies that in a certain region of space different radio access technologies operate using various frequencies. A mobile device can transmit with the access network nodes or with other mobile devices by using licensed bands, unlicensed bands (shared spectrum), through a secondary usage of a dedicated spectrum, or on a spectrum dedicated to cognitive radio systems (CRSs) [13].

An example of the described scenario is presented in Figure 5.20, where the eNBs and NBs transmit on a licensed spectrum, the AP uses unlicensed frequencies and the IEEE 802.22 base station uses TV bands as a secondary user (with its connected devices). In the figure are also represented two cognitive radio networks, one operating in a band assigned to cognitive radio transmissions (long term scenario) and the other operating in an unlicensed shared band (short term scenario).

The cognitive pilot channel supports cognitive terminals both during the startup and ongoing phases. When the mobile device switches on, it searches for the presence of the cognitive pilot channel (CPC) in well-known frequencies. If detected, it synchronizes with the CPC and acquires context information. Then it is able to choose one or more radio networks for association and data transmission. The reconfigurable radio device (RRD) uses CPC to receive information about the available radio access networks in its location, the used frequencies, the operators that own and manage the network, the policies they apply, etc.

Then the terminal registers to one or more access networks and moves to the ongoing phase. During the ongoing phase, the terminal periodically checks the CPC in order to update the previous information and to acquire additional information like offered services, networks capabilities and load, etc. The acquired information is also used to support the decisions of

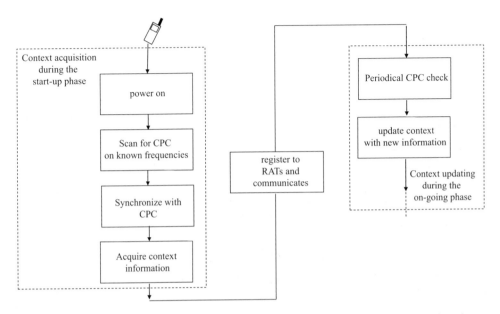

Figure 5.21   Flow chart for context acquisition during the mobile device startup and ongoing phases.

RAT changes (i.e., for load balancing purposes) or of requests for new software downloads (i.e., for installing a new RAT). Figure 5.21 shows the flow chart for context acquisition during the mobile device startup and ongoing phases.

The cognitive pilot channel can be:

- Out-band, if transmitted on a known band dedicated to CPC using a given radio technology.
- In-band, if transmitted in the same band as the RAT network nodes. In-band CPC transmission uses a bidirectional control channel of the selected access network.

5.2.3.1   Out-Band CPC

The out-band cognitive pilot channel (CPC) is a bidirectional channel transmitted on a well-known band and using a well-known radio technology. It broadcasts downlink information about the available radio access technologies (RATs) in a certain geographic region, with the used frequencies from each operating RAT and the available frequencies for primary and secondary spectrum usage, the operators, etc.

The out-band CPC presents two main drawbacks: one is the difficulty of having, all over the world, a unique frequency for CPC transmission; the other is the size of the band assigned to CPC, which is related to the amount of information to be transmitted. On the other hand, an out-band CPC has the evident advantages of being easy and fast to search. In general, an out-band CPC provides startup information to terminals in a geographic area like available radio access networks, operators, used spectrum and policies. The information on a used spectrum is updated with all changes, for example due to dynamic spectrum access (DSA) [14] or flexible

Legacy Mobile Device

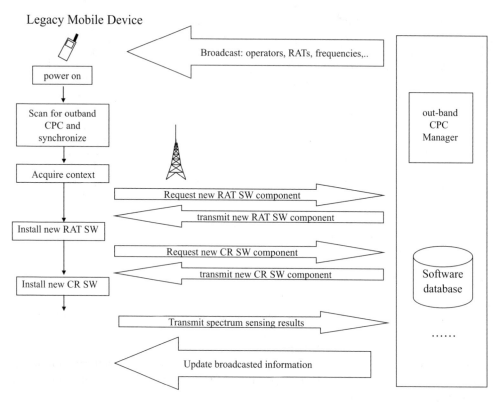

Figure 5.22 Example of out-band CPC usage during the startup phase of a mobile device and for software download.

spectrum management (FSM) [15]. The out-band CPC is in general a bidirectional channel, which in downlink broadcasts context information and in uplink receives measurements from the reconfigurable radio devices (RRDs), like spectrum sensing measurements, etc. The bidirectional out-band CPC can also be used from the RRDs to request new radio applications and new cognitive features (uplink) and to download them (downlink) from a software database [13]. An example of out-band CPC usage during the startup phase of a mobile device and for software download is presented in Figure 5.22.

Out-band CPC can also be used to aid base stations (BSs) and terminals to select the available spectrum for secondary usage and to establish a communication on the chosen white spaces. In this case, when the secondary terminals and base stations are switched on, they scan the well-known frequencies to search for the out-band CPC and synchronize to the downlink in order to obtain information about the available frequencies for secondary usage. Then BSs and terminals sense the spectrum and transmit, on the uplink out-band CPC, the measurement results. When the available frequencies for secondary usage of the spectrum are determined, the secondary network is set up on the chosen set of frequencies and the context is updated. An example of out-band CPC usage for secondary systems assistance during the startup phase is shown in Figure 5.23.

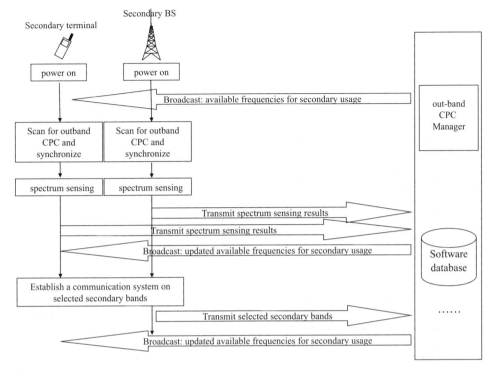

Figure 5.23   Example of out-band CPC usage for secondary systems assistance during the startup phase.

The out-band CPC is also used during the secondary network ongoing phase to exchange information required for distributed decision making. For example, sensing measurement results can be transmitted to the CPC manager to update the available frequencies for secondary usage of the spectrum and eventually change the secondary network operating frequencies.

Examples of Out-Band CPC Deployments
The out-band CPC is transmitted through a well-known radio access technology on well-known frequencies. In [16], two possible solutions are considered.

The first solution deals with the usage of the GSM network for out-band CPC delivery. In this scenario, the GSM base stations must be connected to the out-band CPC manager. The CPC can be mapped into one or more GSM control channels. For example, broadcast information could be transmitted on the broadcast control channel (BCH) and uplink communication could be sent on the random access channel (RACH). For bidirectional communication (e.g., for new software requests and downloads), a dedicated bidirectional control channel like the standalone dedicated control channel (SDCCH) could be used. The advantage of using GSM as the radio access technology for CPC is that it is the radiomobile network with the highest percentage of coverage in Europe and other countries in the world. With that technology, the out-band CPC would be set up with minimum investments. An example of out-band CPC transmission on GSM control channels is presented in Figure 5.24.

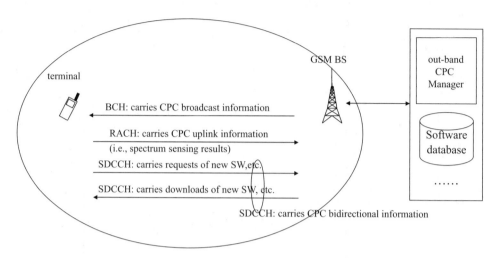

Figure 5.24   Example of out-band CPC transmission on GSM control channels.

The second solution proposed in [13] is the possibility to transmit CPC on unlicensed frequencies like 2.4 GHz using wireless local area networks (WLANs). In this scenario, the access points (APs) must be connected to the CPC manager. CPC broadcast information could be transmitted on the beacon sent from the access point and out-band CPC bidirectional communication could be transmitted on the AP radio channel following the IEEE 802.11 CSMA/CA (carrier sense multiple access with collision avoidance) protocol. Implementation of the out-band CPC on the IEEE 802.11 wireless interface has the drawback of a nonuniform coverage. This would not guarantee the out-band CPC presence everywhere, only in wireless LAN islands. The example of out-band CPC transmission on unlicensed frequencies through the IEEE 802.11 radio interface is represented in Figure 5.25.

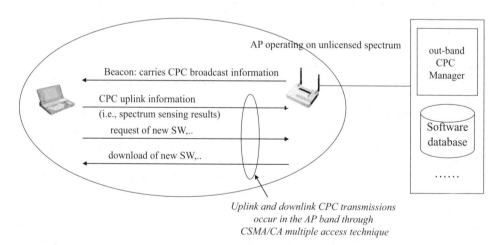

Figure 5.25   Example of out-band CPC transmitted on unlicensed frequencies through the IEEE 802.11 radio interface.

Out-Band CPC Delivery Strategies

A mobile device (MD) can acquire the out-band context through three possible strategies [13]: passive scanning, active scanning and IP-based CPC.

In passive scanning, the MD scans the well-known frequencies for out-band CPC and receives the context information. In this case the context information is continuously broadcast using a new logical channel, the downlink broadcast CPC (DBCPC) channel [13]. The DBCPC is mapped into a broadcast transport channel of the chosen radio access technology.

In active scanning, the MD transmits a request for CPC on a well-known uplink channel, the logical random access CPC (RACPC). The network acknowledges the correct reception of the RACPC with an acquisition indication CPC (AICPC) and transmits the CPC information on the downlink on demand CPC (DODCPC). RACPC and DODCPC are mapped into uplink and downlink transport channels of the chosen radio access technology; the AICPC is an indicator added at the physical layer.

Figure 5.26 shows an example where CPC channels are added to GSM radio channels. In the figure, DBCPC, RACPC and DODCPC are new logical channels mapped into existing transport channels, while AICPC is a new indicator at the physical layer. Figure 5.27 shows the messages exchanged between the terminal and network for passive and active CPC scanning.

Other than passive and active CPC scanning, a third delivery strategy has been proposed in [13], which is the IP-based CPC. In this case, the CPC implementation is at the application layer and the network architecture is provided by a CPC information database reachable through an Internet connection. When switched on, the mobile device connects to one available RAT to establish an Internet connection with the CPC information database. Then it requests CPC information and, after the context reception, eventually decides to move to other RATs. An example of the IP-based CPC architecture is shown in Figure 5.28. The advantage of the IP-based CPC architecture is that the out-band CPC manager can be wherever it is connected to the Internet and can be managed from operators or third parties.

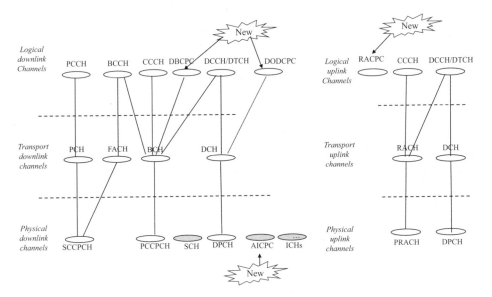

Figure 5.26   Example of CPC channels added to the GSM radio channel structure.

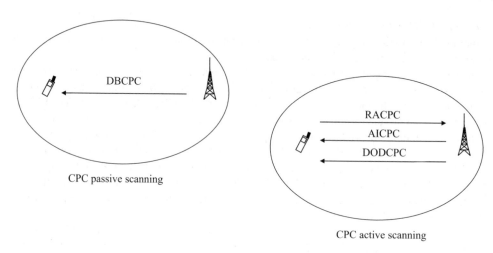

Figure 5.27   Example of passive and active CPC scanning.

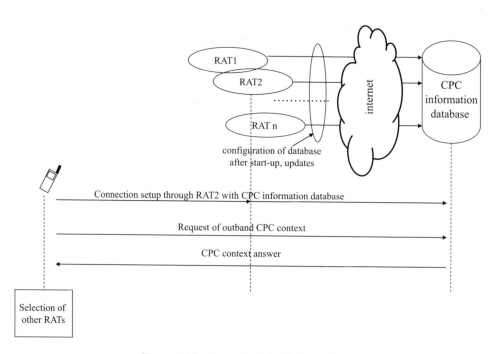

Figure 5.28   Example of the IP-based CPC.

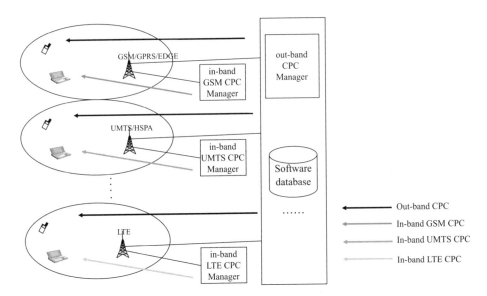

Figure 5.29   Example of network architecture with out-band and in-band CPCs.

5.2.3.2   In-band CPC

The in-band CPC is sent on control channels of the same radio access technology used for data transmission. In-band CPC, rather than an alternative, is a complement to the out-band CPC. In the combined CPC architecture, the out-band CPC broadcasts the same information over the whole region. The out-band CPC broadcasted information is about radio access networks, operators, used spectrum and policies. Each operating radio access network is provided with an in-band CPC with additional information like RAT type, operator code, policies, capabilities, etc. An example of network architecture with out-band and in-band CPCs is presented in Figure 5.29.

When the mobile device switches on, it listens for the out-band CPC in order to receive global information about the context, like the available radio access technologies (RATs) in that location, the frequencies in use and the available frequencies for secondary usage. The obtained information is used to select and connect to one or more RATs. Then the device stops listening to the out-band CPC and starts to listen to the in-band CPC, in order to receive the context associated with the chosen RATs. If the mobile device needs to move the actual connection to a different RAT, it goes to listen again for the out-band CPC. The flow chart showing the described out-band and in-band CPC acquisition is shown in Figure 5.30.

In-band CPC delivery strategies can be similar to out-band CPC delivery strategies. The in-band CPC can be mapped into control channels of the given radio access technology, with passive or active modes of context acquisition. In-band CPC can also be implemented based on an IP connection to a CPC server.

Use Cases for In-Band CPC

The in-band CPC is used to deliver CPC context information in the same band of data transmission. In the scenario where multiple RATs are managed from the same network

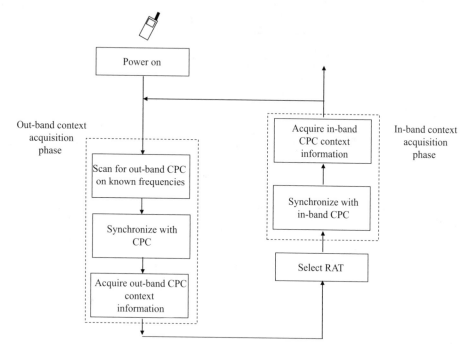

Figure 5.30   Flow chart for out-band and in-band CPC acquisition.

operator, in-band CPC can also deliver instructions to the mobile devices to perform network optimization [17].

The entities involved in the given scenario are: mobile devices, multiple radio access technologies managed from the same network operator, joint radio resource management (JRRM) entity, configuration control module (CCM) and the CPC manager. The access network nodes are reconfigurable nodes equipped with hardware processing resources (HPR) shared among different RATs. Each RAT is a radio application using a certain amount of HPRs depending of the configuration of radio parameters, the amount of ongoing traffic, etc. The HPR is connected to the radio frequency block (RFB), which goes to the antenna and is able to support different RATs at the same time.

The JRRM entity handles the radio resources, performing radio bearer control, radio admission control, scheduling of resources in uplink and downlink, load balancing, etc. The JRRM entity can be implemented inside the RBS (as represented in figure 5.5) or as an external entity.

Another entity in the scenario considered for the use case is the reconfiguration entity (RE), which can be internal or external to the RBS and has the task of evaluating possible RBS reconfigurations.

In the described use case, RBS, JRRM and the reconfiguration entity are three distinct entities. Finally, there is the in-band CPC manager handling the in-band CPC. Figure 5.31 shows an example of in-band CPC usage for load balancing (RAT reselection).

After power on, the terminal camps on RAT1. The output of the load balancing algorithm of the radio resource management (RRM) gives RAT2 as preferred. This information is

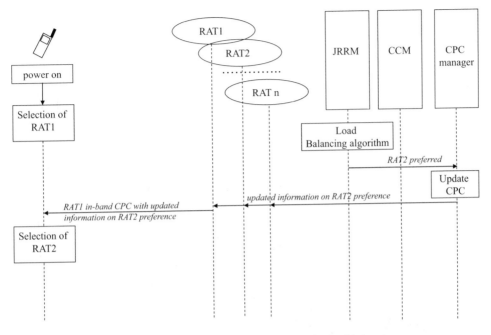

Figure 5.31   Example of in-band CPC usage for load balancing.

communicated from the RRM entity to the CPC manager, which updates CPC information with the RAT2 preference. This updated preference is sent on the RAT1 in-band CPC, which is periodically checked from the mobile device. After reception of the new policy, the terminal moves to RAT2, thus ending the CPC-aided load balancing procedure. A similar procedure can be run during the ongoing phase for radio resource optimization. Figure 5.32 shows an example of the CPC-aided radio resource optimization procedure with RBS software reconfiguration.

During the ongoing phase, the terminal uses RAT2 for data transmission. The output of the load radio resource optimization algorithm of RRM gives RAT3 as preferred. This information is communicated from the RRM entity to the CPC manager, which updates the CPC information with the RAT3 preference. Because the RBS is not provided with RAT3 software, it downloads it from the CCM, which in this example includes the software database. It then updates the in-band CPC with the information that RAT3 is preferred. After reception of the new policy, the terminal moves to RAT3, thus ending the CPC-aided radio resource optimization procedure.

5.2.4   ETSI RRS Functional Architecture

In 2009, the European Telecommunication Standard Institute (ETSI) proposed a functional architecture (FA) for the management and control of reconfigurable radio systems (RRSs) [3]. The scenario is of different RATs connected to an IP-based core network, as represented in Figure 5.2. The different RATs may belong to the same or to different operators, and include different types of access network nodes, like legacy GSM base stations (BSs), UMTS NBs,

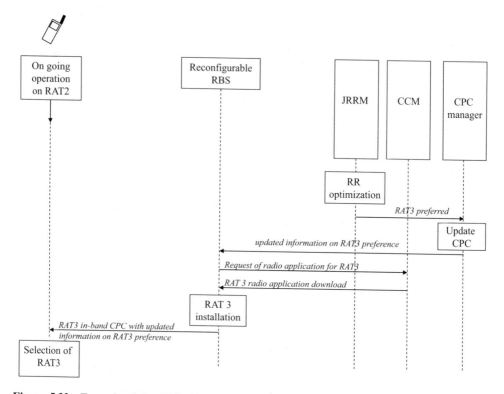

Figure 5.32 Example of the CPC-aided radio resource optimization procedure with RBS software reconfiguration.

LTE eNBs, multi-RAT reconfigurable RBSs, cognitive radio BSs, access points, etc. Then, existing and new RATs are supposed to coexist and be integrated through an architecture that will globally and dynamically optimize the radio resources according to spectrum policies and user needs.

Different types of radio terminals operate in this multi-RAT scenario, like legacy 2G/3G terminals, cognitive devices and multi-RAT reconfigurable radio devices. The terminals can communicate with the access network nodes or with direct links. Some of them also support multiple connections with different RATs. The connections respect service-dependent QoS parameters and other policies related, for example, to the user profile, the used RATs, etc.

The access network nodes are connected to several servers for the management of the reconfigurable nodes. Some of the servers give the spectrum policies and also flexibly manage the spectrum allocated to the RATs, even in the case where they belong to different operators. Spectrum sharing and spectrum renting can be allowed or not, and primary and secondary spectrum usage can be set in some parts of the band with specified rules. The rules for spectrum usage can be changed according to the regulator policies [10].

Figures 5.33, 5.34 and 5.35 show the high level ETSI functional architecture for the management and control of reconfigurable radio systems in three cases: single operator point of view, multioperator point of view and single operator multihop point of view.

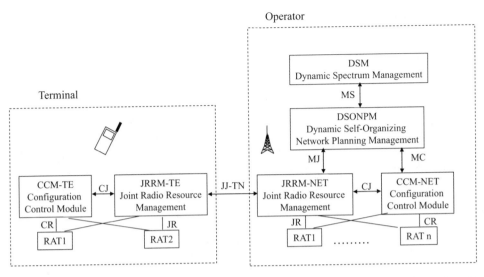

Figure 5.33 High level ETSI functional architecture for the management and control of reconfigurable radio systems in the case of a single operator point of view. (Reproduced with permission. Copyright European Telecommunications Standards Institute 2008. Further use, modification, copy and/or distribution are strictly prohibited. ETSI standards are available from http://pda.etsi.org/pda/.)

In Figure 5.33 the ETSI FA is represented and shows the case of a terminal connected to an operator. Figure 5.34 shows the same FA in the case of a multioperator scenario, while in Figure 5.35 the case of a multihop architecture is presented. In the multioperator scenario (Figure 5.34), a terminal can be connected to one or more network operators (NOs) and the NOs cooperate to optimize the whole radio resource usage. In multihop architecture (Figure 5.35), terminal 1 can directly communicate with the network, can operate as the relay of terminal 2 and can also directly communicate with terminal 2.

The operator functional architecture includes the following functional blocks:

- DSM (dynamic spectrum management)
- DSONPM (dynamic self-organizing network planning management)
- CRRM-NET (configuration control module – network side)
- JRRM-NET (joint radio resource management – network side)

The terminal functional architecture includes the following blocks:

- CCM-TE (configuration control module – terminal equipment side)
- JRRM-TE (joint radio resource management – terminal equipment side)

The configuration control module (CCM) communicates with the JRRM block both at the terminal and operator sides through the standard CJ interface. CCM controls the configuration of the different RATs through the CR interface and JRRM jointly manages the radio resources of the different RATs through the JR interface, in both network and terminal sides. The JRRM

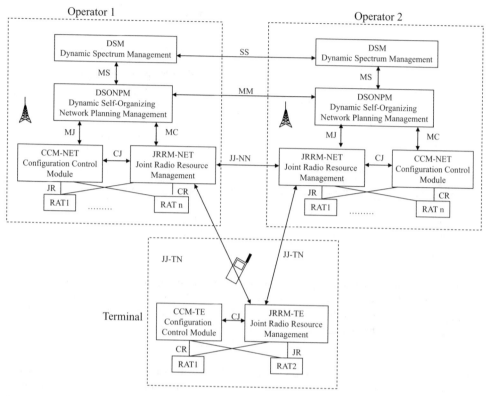

Figure 5.34 High level ETSI functional architecture for the management and control of reconfigurable radio systems in the case of a multioperator point of view. (Reproduced with permission. Copyright European Telecommunications Standards Institute 2008. Further use, modification, copy and/or distribution are strictly prohibited. ETSI standards are available from http://pda.etsi.org/pda/.)

modules of the terminal and the network communicate through a standard interface named JJ-TN (JJ-terminal–network).

The self-organizing module of the operator network, the DSONPM, communicates with the CCM-NET and the JRRM-NET modules respectively through the MJ and MC interface. Finally, the DSM module of the operator dynamically manages the spectrum through a bidirectional communication with the DSONPM enabled through the MS interface.

If there are two operators (Figure 5.34), the interfaces between the functional blocks enable the joint optimization of the whole radio resources. In particular, SS is the interface between DSM of operator 1 and DSM of operator 2; MM is the interface between DSONPM of operator 1 and DSONPM of operator 2; JJ-NN (JJ-network–network) is the interface between the JRRM modules of the two operators. In the case of terminal–terminal communication, the JJ-TT (JJ-terminal–terminal) interface enables the communication between the JRRM modules of the two devices.

Messages between DSONPM and CCM are exchanged through the MC interface; messages between DSONPM and JRRM are exchanged through the CJ interface.

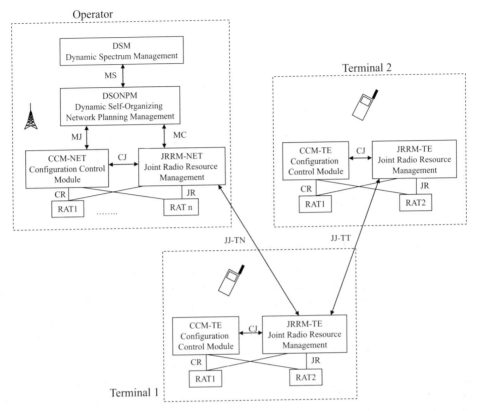

Figure 5.35 High level ETSI functional architecture for the management and control of reconfigurable radio systems in the case of a single operator multihop point of view. (Reproduced with permission. Copyright European Telecommunications Standards Institute 2008. Further use, modification, copy and/or distribution are strictly prohibited. ETSI standards are available from http://pda.etsi.org/pda/.)

5.2.4.1 Dynamic Spectrum Management Entity

The DSM is the entity that manages the operator spectrum resources. It enforces spectrum policies imposed by the regulator and chosen from the operator (i.e., enabling of spectrum trading). It knows the current spectrum assignments, including primary and secondary spectrum usages. It provides the bandwidth allocations to the DSONPM, which maintains the updated context and gives to each RAT the available amount of spectrum. Moreover, if available, it trades the spectrum with the other operators. The DSM is aided in the decision making process by the information coming from the DSONPM entity, which provides to the DSM indicators, like spectral efficiency, spatial spectrum usage, etc. Messages between DSM and DSONPM are exchanged through the MS interface. If spectrum trading is enabled, the SS interface among different operators allows the possibility of dynamic spectrum renting and offering. Figure 5.36 shows the DSM functionalities with interfaces.

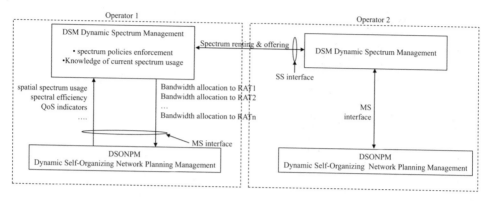

Figure 5.36 DSM functionalities with interfaces.

5.2.4.2 Dynamic Self-Organizing Network Planning Management Entity

The DSONPM is the entity that plans the medium and long term reconfiguration actions that the multi-RAT network should undertake. The DSONPM acquires information about the context from the JRRM module, regarding the network elements of the different RATs (i.e., resources used in each RAT and cell, operating frequencies, the traffic that each RAT is handling with offered QoS levels, service priorities, etc.). It also acquires from the JRRM information about the mobile devices, like used resources and frequencies, etc.

The DSONPM receives from the CCM the current configuration of RAT network elements other than their reconfiguration capabilities. Moreover, it acquires from the CCM information about the mobile devices capabilities, their current configuration and reconfiguration capabilities, etc. The DSONPM then runs a self-optimization process that leads to reconfiguration actions towards the RAT network nodes. The decisions taken from the DSONPM also consider the network operator policies concerning optimization and negotiation algorithms, criteria regarding user categories, QoS, preferences of RATs for certain services, etc. The output of the DSONPM module gives inputs to the JRRM and CCM entities. To the CCM, it requests new reconfigurations of certain RATs, cells and mobile devices giving the new configuration parameters; to the JRRM, it requests changes about the resources used in RATs, cells, QoS levels, modifies the resources assigned to some users, etc.

If the cooperation between network operators is enabled, the interface MM can be used to exchange information about network configuration and planning in order to avoid interference and to jointly optimize network performances. Figure 5.37 shows the DSONPM functionalities with interfaces.

5.2.4.3 Joint Radio Resource Management Entity

The joint radio resource management (JRRM) is the entity that jointly manages the radio resources of the network and of the terminals. It is divided into two entities: JRRM-NET, which is the instance on the network side, and the JRRM-TE, which is the instance on the terminal side. On the network side, a JRRM instance could be active on each base station or it can be centralized in one or more controllers of the radio resources.

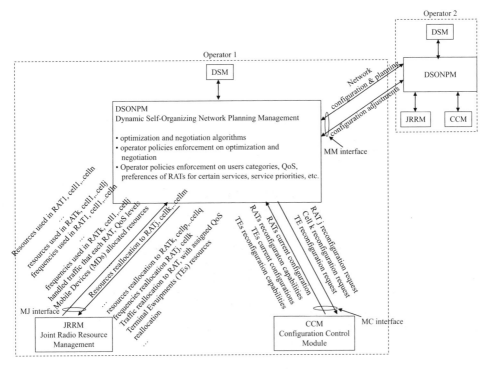

Figure 5.37   DSONPM functionalities with interfaces.

The JRRM-TE is responsible for access selection of a terminal in idle mode. The access selection of a terminal is jointly managed from JRRM on both the network and terminal sides and depends on terminal capabilities, requested services, user preferences, current network load and radio channel estimates. The access selection includes the choice of the operator, the RAT, and the cell. The JRRM entity also runs all the existing and new algorithms for the optimal management of radio resources, such as the radio bearer control, radio admission control, scheduling of resources in uplink and downlink, retransmission handling, etc.

JRRM-NET and JRRM-TE communicate through the standard interface JJ-TN. The JJ-TN interface is used from the network to send context information, such as which RATs/cells are available in a certain area, which are preferred, the policies for access selection, cells location and capabilities, etc. For example, the cognitive pilot channel can be mapped on the JJ-TN interface. The network also uses the JJ-TN interface to command handover decisions, resources reallocation on active links, etc. In the other direction, the JJ-TN interface is used from the mobile devices to send measurement results, spectrum sensing results, performances on active links, etc. In a communication between two terminals (i.e., a direct link), the JRRM-TE instances of the involved terminals exchange information about the context (i.e., measurements results, spectrum sensing results) and negotiate the resources through the JJ-TT interface. In a multihop configuration, the JJ-TT interface is also used to forward information from and to the JJ-TN interface.

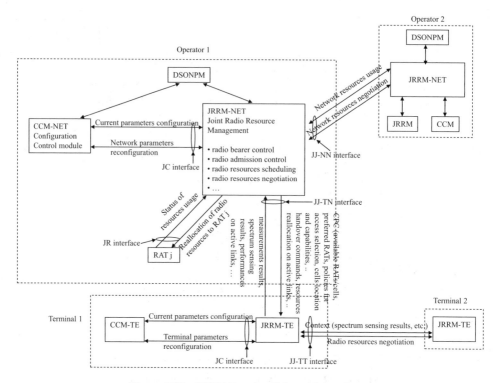

Figure 5.38   JRRM functionalities with interfaces.

The JRRM module communicates with the configuration control module (CCM) through the CJ interface, for the exchange of current and new configuration parameters. This applies in both network and terminal sides.

Finally, the JRRM communicates with the RATs through the JR interface, used for reporting resource usage status and for reallocations of radio resources to the RATs. The JRRM can also request certain RAT measurements of some radio parameters and the related reporting. The ETSI technical report [1] recommends that the measurements be described in a general format (i.e., bit rate, bit error rate, delay) and RAT-related parameters or terminal class related parameters (i.e., channel quality indicator) should not be used. The communication of the JRRM entity with the underlying RATs applies on both network and terminal sides.

If the cooperation between network operators is enabled, the interface JJ-NN can be used to exchange information about network resource usage and negotiation in order to avoid interference and to jointly optimize network performances. Figure 5.38 shows the JRRM functionalities with the related interfaces.

5.2.4.4   Configuration Control Module

The configuration control module (CCM) is the entity that is responsible for the configuration of network elements. It is divided into two entities: CCM-NET, which is the instance on the

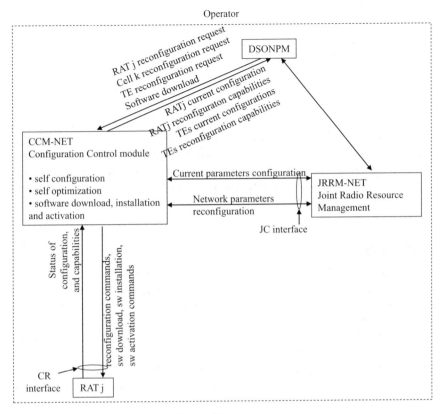

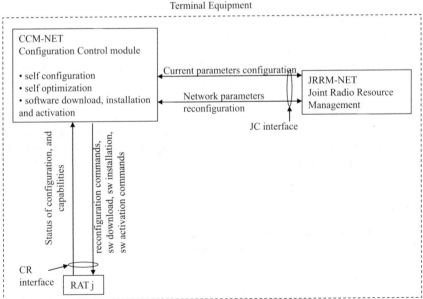

Figure 5.39 CCM functionalities with interfaces.

network side, and CCM-TE, which is the instance on the terminal side. On the network side, a CCM instance could be active in each base station or in general in each reconfigurable node. The CCM-TE is responsible for the terminal equipment reconfiguration.

The CCM executes the directives received from the JRRM through the JC interface and from the DSONPM through the MC interface. It performs self-configuration and self-optimization of reconfigurable network nodes. For example, in a base station, it can switch on/off some carriers and modify antenna tilts, etc. On both terminal and network sides, it switches some sessions to other cells or RATs, and is also responsible for software (i.e., new RATs or new features of existing RATs) download, installation and activation. CCM reports information on the current configuration to the JRRM module, like the availability of new software and the corresponding configuration. The CCM module receives from the underlying RATs the current configuration of radio applications and requests the execution of reconfigurations. Figure 5.39 shows the CCM functionalities with the related interfaces.

5.3 Summary

In 2009, the European Telecommunication Standard Institute published the first technical reports on reconfigurable radio systems. ETSI proposes a reference architecture for a reconfigurable radio base station with multistandard capabilities, with different RATs implemented as software and the possibility to download, install and activate new RATs or update existing RATs with new features. The RBS also includes self-establishment and self-optimization functionalities, other than the possibility of flexible spectrum allocation.

The reference architecture for a reconfigurable mobile device implements the terminal as a radio computer, with radio applications for different RATs. Multiple radios can be run simultaneously in the radio computer, with the possibility to reconfigure and adjust radio parameters.

The ETSI recommendations for standardization include the possibility to convey the radio context (i.e., used RATs, available frequencies for primary and secondary usage, used frequencies, operator policies, etc.) in a cognitive pilot channel (CPC), which can be implemented in the same band of the current RAT (in-band CPC) or in a dedicated band (out-band CPC). The proposed functional architecture envisages scenarios with a single operator and different network operators collaborating for the joint goal of radio resources optimization.

New entities are introduced for dynamic spectrum management, dynamic self-organizing network planning and management, joint radio resource management and configuration control. The new entities bidirectionally communicate through standard interfaces and with their corresponding entity in the radio terminal.

References

1. ETSI TR 102 802 (2009) *Reconfigurable Radio Systems (RRS); Cognitive Radio System Concept*, v1.1.1.
2. ETSI TR 102 733 (2010) *Reconfigurable Radio Systems (RRS); System Aspects for Public Safety*, V1.1.1.
3. ETSI TR 102 682 (2009) *Reconfigurable Radio Systems (RRS); Functional Architecture for Management and Control of Reconfigurable Radio Systems*, v1.1.1.
4. Mueck, M., Piipponen, A., Kalliojärvi, K. *et al.* (2010) ETSI Reconfigurable Radio Systems: Status and Future Directions on Software Defined Radio and Cognitive Radio Standards. *IEEE Communications Magazine*, September, 78–86.

5. ETSI TR 102 681 (2009) *Reconfigurable Radio Systems (RRS); Radio Base Station (RBS) Software Defined Radio (SDR) Status, Implementations, and Costs Aspects, Including Future Possibilities*, v1.1.1.
6. Open Base Station Architecture Initiative, BTS System Reference Document, version 2.0, 2006.
7. ETSI TR 102 680 (2009) *Reconfigurable Radio Systems (RRS); SDR Reference Architecture for Mobile Device*, v1.1.1.
8. ETSI TR 102 628 (2010) *Electromagnetic Compatibility and Radio Spectrum Matters (ERM); System Reference Document; Land Mobile Service; Additional Spectrum Requirements for Future Public Safety and Security (PSS) Wireless Communication Systems in the UHF Frequency Range*, V1.1.1.
9. ETSI TR 102 839 (2011) *Reconfigurable Radio Systems (RRS); Multiradio Interface for Software Defined Radio (SDR) Mobile Device Architecture and Services*, V1.1.1.
10. ETSI TR 102 803 (2010) *Reconfigurable Radio Systems (RRS); Potential Regulatory Aspects of Cognitive Radio and Software Defined Radio Systems*, V1.1.1.
11. ETSI TR 103 064 (2011) *Reconfigurable Radio Systems (RRS); Business and Cost Considerations of Software Defined Radio (SDR) and Cognitive Radio (CR) in the Public Safety Domain*, V1.1.1.
12. ETSI TR 103 062 (2011) *Reconfigurable Radio Systems (RRS); Use Cases and Scenarios for Software Defined Radio (SDR) Reference Architecture for Mobile Device*, V1.1.1.
13. ETSI TR 102 802 (2010) *Reconfigurable Radio Systems (RRS); Cognitive Radio System Concept*, V1.1.1.
14. Clancy III, T.C. (2006) Dynamic Spectrum Access in Cognitive Radio Networks, Dissertation submitted to the Faculty of the Graduate School of the University of Maryland, College Park, in partial fulfillment of the requirements for the degree of Doctor of Philosophy.
15. Roorda, P. (2011) Critical Issues for the Flexible Spectrum Network, JDSU White Paper, April.
16. ETSI TR 102 683 (2009) *Reconfigurable Radio Systems (RRS); Cognitive Pilot Channel (CPC)*, v1.1.1.
17. ETSI TR 103 063 (2011) *Reconfigurable Radio Systems (RRS); Use Cases for Reconfigurable Radio Systems Operating in IMT Bands and GSM Bands for Intra-operator Scenarios*, V1.1.1.

6

IEEE 1900.4

6.1 Introduction

The access network evolution has led in recent years to scenarios in which a single IP-based backbone provides the transport service to multiple access networks, both fixed and wireless. The development of cognitive radio terminals and networks will introduce new scenarios in the radio access network.

With spectrum sensing cognitive radio terminals and networks, spectrum usage will be optimized among different radio access networks (RANs). The available spectrum will be shared among different RANs depending on parameters like location, time, actual policies, agreements among the operators, etc. Spectrum sharing will also introduce a spectrum renting opportunity, with the possibility to offer opportunistically the unused spectrum for renting.

The IEEE standard 1900.4 on 'Architectural building blocks enabling network-device distributed decision making for optimized radio resource usage in heterogeneous wireless access networks, came out in 2009 and is an IEEE standard for reconfigurable radio systems (RRS) [1]. The context is a heterogeneous environment with multiple operators and multiple radio access networks (RANs). In this scenario, an IP-based backbone provides the transport service to the multiple RANs. IEEE 1900.4 standardizes the overall system architecture for reconfigurable radio networks and terminals, with the goal of overall radio resources optimization while maintaining the requested quality of service (QoS).

In 2011, an amendment to the IEEE 1900.4 standard, the IEEE 1900.4a, was published. IEEE 1900.4a adds to the IEEE 1900.4 functional architecture new entities for dynamic spectrum access networks in white space frequency bands.

The chapter describes the IEEE 1900.4 and IEEE 1900.4a functional architectures, with the functional entities that enable reconfigurable radio system architectures. For both IEEE 1900.4 and IEEE 1900.4a architectures examples and use cases are presented. These come from the author and not from the standard and are intended to provide high level examples of operation of the functional architecture.

Reconfigurable Radio Systems: Network Architectures and Standards, First Edition. Maria Stella Iacobucci.
© 2013 John Wiley & Sons, Ltd. Published 2013 by John Wiley & Sons, Ltd.

6.2   IEEE Dynamic Spectrum Access Networks Standards Committee (DySPAN-SC)

In 2005 the IEEE P1900 Standard Committee was established to develop new standards to improve the use of spectrum. Starting from 2005, IEEE began to support projects related to cognitive radio (CR), the IEEE 1900P series.

In 2007, the IEEE 1900 was reorganized as the IEEE Standard Coordination Committee 41 (IEEE SCC41), created with the goal of developing standards related to dynamic spectrum access networks. To reach this goal, the new standards had to introduce new techniques able to manage interference, share the information, coordinate the wireless technologies and manage a multiradio access network (multi-RAN) [2]. The IEEE Communication Society and IEEE Electromagnetic Compatibility Society supported IEEE 1900 and continued to support SCC41.

In December 2010, the IEEE SCC41 was reorganized as the IEEE Dynamic Spectrum Access Networks Standards Committee (DySPAN-SC) and its sponsor was changed to the IEEE Communications Society Standards Development Board (CSDB) [3]. Figure 6.1 shows the history of the actual DySPAN-SC.

The IEEE DySPAN-SC includes the following Working Groups (WGs) [3]:

- IEEE 1900.1: Working Group on Definitions and Concepts for Dynamic Spectrum Access: Terminology Relating to Emerging Wireless Networks, System Functionality, and Spectrum Management. The IEEE 1900.1 standard was published in 2008. From 2 February 2011, the 1900.1 Working Group works on a new project:
  - 1900.1a: IEEE Standard Definitions and Concepts for Dynamic Spectrum Access: Terminology Relating to Emerging Wireless Networks, System Functionality, and Spectrum Management. Amendment: Addition of New Terms and Associated Definitions.
- IEEE 1900.2: Working Group on Recommended Practice for the Analysis of In-Band and Adjacent Band Interference and Coexistence between Radio Systems. The IEEE 1900.1 standard was published in 2008 and provides guidance for the analysis of coexistence and interference among various radio access technologies scenarios.
- IEEE 1900.3: Working Group on Recommended Practice for Conformance Evaluation of Software Defined Radio (SDR) Software Modules. This WG has been disbanded.

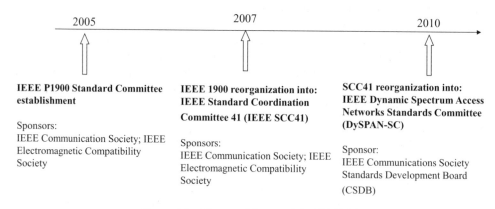

Figure 6.1   History of the actual DySPAN-SC.

- IEEE 1900.4 Working Group on Architectural Building Blocks Enabling Network-Device Distributed Decision Making for Optimized Radio Resource Usage in Heterogeneous Wireless Access Networks. This standard was published in February 2009 and standardizes a functional architecture for reconfigurable radio network and terminals, enabling the optimization of usage of radio resources in a scenario with heterogeneous access networks connected to a unique packet-based core. From April 2009, 1900.4 Working Group works on two projects:
  - 1900.4a: Standard for Architectural Building Blocks Enabling Network-Device Distributed Decision Making for Optimized Radio Resource Usage in Heterogeneous Wireless Access Networks – Amendment: Architecture and Interfaces for Dynamic Spectrum Access Networks in White Space Frequency Bands.
  - 1900.4.1: Standard for Interfaces and Protocols Enabling Distributed Decision Making for Optimized Radio Resource Usage in Heterogeneous Wireless Networks.
- IEEE 1900.5 Working Group on Policy Language and Policy Architectures for Managing Cognitive Radio for Dynamic Spectrum Access Applications. It defines a vendor-independent set of policy-based control architectures and corresponding policy language requirements for managing the functionality and behaviour of dynamic spectrum access networks [3]. 1900.5 WG works on two other projects:
  - IEEE 1900.5a: provides an amendment to P1900.5 defining the interface description between policy architecture components.
  - IEEE 1900.5.1: defines a vendor-independent policy language for managing the functionality and behaviour of dynamic spectrum access networks based on the language requirements defined in the IEEE 1900.5 standard.
- IEEE 1900.6 Working Group on Spectrum Sensing Interfaces and Data Structures for Dynamic Spectrum Access and Other Advanced Radio Communication Systems. This standard was published in April 2011 and defines the interfaces and data structures required to exchange sensing-related information [4]. From June 2011, the 1900.6 Working Group has been working on the 1900.6a project, which is an amendment to 1900.6 adding procedures, protocols and message format specifications for the exchange of sensing related data, control data and configuration data between spectrum sensors and their clients. It also specifies the interfaces between the data archives and other data sources.
  - 1900.6a: Standard for Spectrum Sensing Interfaces and Data Structures for Dynamic Spectrum Access and Other Advanced Radio Communication Systems. Amendment: Procedures, Protocols and Data Archive Enhanced Interfaces.
- IEEE 1900.7: White Space Radio Working Group. This is under development and deals with Radio Interface for White Space Dynamic Spectrum Access Radio Systems Supporting Fixed and Mobile Operation [3].

Figure 6.2 shows the IEEE DySPAN Standards Committee structure.

6.3 IEEE 1900.4 Functional Architecture

In February 2009, the IEEE 1900.4 standard published a functional architecture for reconfigurable radio networks and terminals, enabling the optimization of usage of radio resources in a scenario with heterogeneous access networks connected to a unique packet

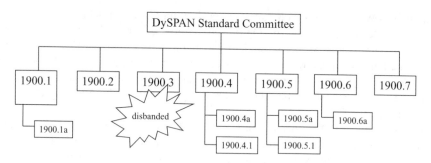

Figure 6.2   IEEE DySPAN Standards Committee structure.

based core, as represented in Figure 6.3. The different radio access networks may belong to the same or to different operators, and include different types of access network nodes, like legacy GSM base stations (BSs), UMTS NBs, LTE eNBs, multiradio access technology (multi-RAT) reconfigurable radio base stations (RBSs), cognitive radio BSs, access points (APs), etc. Existing and new radio access networks (RANs) are supposed to coexist and be integrated through an architecture that will globally and dynamically optimize the radio resources according to spectrum policies and user needs. This multiradio access network is named in IEEE 1900.4 the composite wireless network (CWN).

Different types of radio terminals operate in this multi-RAN scenario, such as legacy 2G/3G terminals, cognitive devices and multi-RAT reconfigurable radio devices. The terminals can directly communicate with the access network nodes or with direct links. Some of them also support multiple connections with different radio access networks (RANs). The connections respect service-dependent QoS parameters, and other policies related, for example, to the user profile, the used RAN, etc.

The access network nodes are connected to several servers for the management of the reconfigurable nodes. Some of the servers assign the spectrum policies and also flexibly

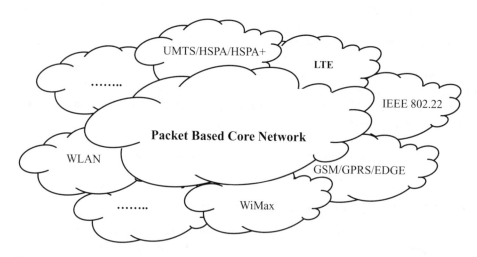

Figure 6.3   Heterogeneous access networks scenario.

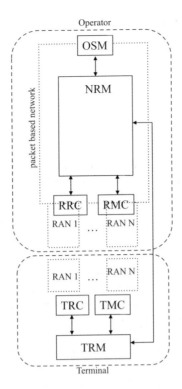

Figure 6.4 High level IEEE 1900.4 functional architecture for the management and control of reconfigurable radio systems in the case of a single operator scenario.

manage the spectrum allocated to the RANs, even in the case where they belong to different operators. Spectrum sharing and spectrum renting can be allowed or not, and primary and secondary spectrum usage can be set in some parts of the band with specified rules. The rules for spectrum usage can be changed according to the regulator policies [5, 6].

Figure 6.4 shows the IEEE 1900.4 functional architecture for reconfigurable radio systems in the case of a single operator scenario [1]. The operator functional architecture includes the following functional blocks:

- Operator spectrum manager (OSM): is the entity that manages the operator spectrum resources. It enforces spectrum policies imposed by the regulator and chosen from the operator. It knows the current spectrum assignments, including primary and secondary spectrum usages. Moreover, if available, it trades the spectrum with the other operators. The OSM is aided in the decision making process by the information coming from the NRM, such as RAN reconfiguration decisions.
- Network reconfiguration manager (NRM): it manages the composite wireless network and terminals to jointly optimize the radio resources of the radio access networks while guarantying the required quality of service (QoS). The NRM obtains spectrum assignment policies from the OSM and receives the context information from the terminal reconfiguration manager (TRM) and from the RAN measurement collector (RMC). The NRM generates and

sends to the TRM radio resource selection policies, which is the framework within which the terminal creates the terminal reconfiguration decisions. The NRM also sends resource selection policies to the RAN reconfiguration controller (RRC). In a multioperator scenario, context information, spectrum assignment policies, radio resource selection policies and reconfiguration decisions are exchanged among different NRMs for joint management and optimization of spectrum and radio resources. The case of the multioperator scenario is shown later in Figure 6.6.

- RAN measurement collector (RMC): it collects the radio access networks context information and sends it to the NRM.
- RAN reconfiguration controller (RRC): is the entity that controls the reconfiguration of different RANs based on the resource selection policies coming from the NRM.

The terminal functional architecture includes the following blocks:

- Terminal reconfiguration manager (TRM): it manages the terminal reconfiguration. It communicates with the network reconfiguration manager (NRM) in order to optimize the radio resources, including spectrum usage. It decides the terminal reconfiguration and selects the most appropriate RAN. To do that, it receives context information from the terminal measurement collector (TMC).
- Terminal reconfiguration controller (TRC): it and controls the reconfiguration on the basis of the requests coming from the TRM.
- Terminal measurement collector (TMC): it collects measurements to obtain context information. The TMC provides context information to the TRM.

Figure 6.5 shows an example of information exchange among operator and terminal functional blocks.

If there are two operators, the communication between the functional blocks of the operators enables the joint optimization of the whole radio resources. Figure 6.6 shows the case of two operators with communication between the corresponding network reconfiguration managers.

6.3.1   Operator Spectrum Manager Entity

The OSM is the entity that manages the operator spectrum resources. It enforces spectrum policies imposed by the regulator and chosen from the operator and, if enabled, trades the spectrum with other operators. The OSM controls the spectrum assignment to the different RANs from the network reconfiguration manager (NRM). It receives RAN reconfiguration decisions from the NRM and transmits spectrum assignment policies.

Figure 6.7 shows the OSM functionalities with interfaces. In Figure 6.7 the OSMs of two operators are represented, showing that spectrum trading among operators is handled by the OSMs through the communication between the NRMs. In fact, in the IEEE 1900.4 architecture the interface OSM–OSM is not provided.

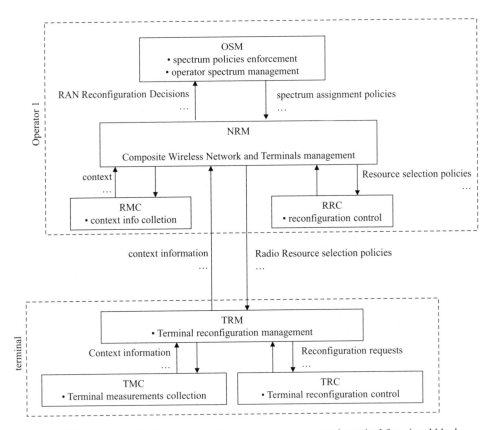

Figure 6.5 Example of information exchange among operator and terminal functional blocks.

6.3.2 Network Reconfiguration Manager Entity

The network reconfiguration manager (NRM) is the entity that assigns the spectrum to different RANs and manages the reconfigurations of the radio access networks and terminals in order to optimize the whole radio parameters. It acquires information about the context from the RAN measurement collectors (RMCs) regarding the network elements of the different RANs (i.e., resources used in each RAN and cell, operating frequencies, the traffic that each RAN is handling with offered QoS levels, etc.). It also acquires information and measurement results from the TRMs of the terminals (i.e., used resources and frequencies, capabilities, current configuration and reconfiguration capabilities, etc.). The NRM receives from the RAN reconfiguration controller (RRC) the current configuration of RAN network elements other than their reconfiguration capabilities.

The NRM then runs a self-optimization process that leads to reconfiguration actions towards the RAN nodes. The decisions taken from the NRM also consider the network operator policies concerning optimization and negotiation algorithms, criteria regarding users categories, QoS, preferences of RANs for certain services, etc. The output of the NRM module gives input to the RMC, the RRC and the TRMs of the mobile devices. To the RRC, it requests new reconfigurations of certain RANs; to the RMC, it gives the new radio parameters (i.e.,

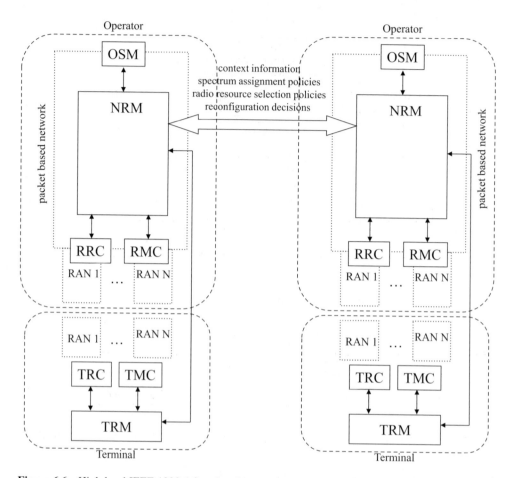

Figure 6.6 High level IEEE 1900.4 functional architecture for the management and control of reconfigurable radio systems in the case of a multiple operator scenario.

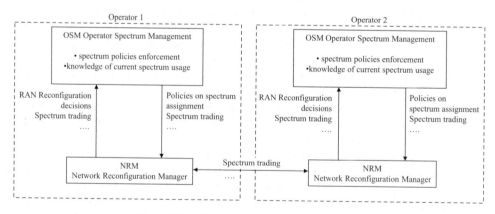

Figure 6.7 OSM functionalities with interfaces.

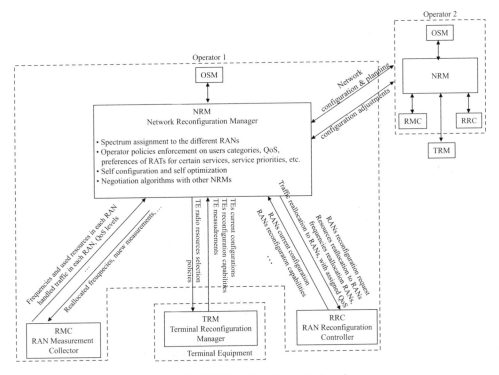

Figure 6.8 NRM functionalities with interfaces.

reassigned frequencies) in order to perform correct measurements; to the TRMs, it gives radio resource selection policies to guide the terminals in their reconfiguration.

If cooperation between network operators is enabled, the interface between NRMs can be used to exchange information about network configuration and planning in order to avoid interference and to jointly optimize network performances. Figure 6.8 shows the NRM functionalities with interfaces.

6.3.3 RAN Reconfiguration Controller and RAN Measurement Collector Entities

The RAN reconfiguration controller (RRC) is the entity that is responsible for the configuration of RAN elements. It can be centralized in a unique node or it can be distributed, that is an RRC instance can be active in each base station or in general in each reconfigurable node. The RRC receives the directives from the network reconfiguration manager and reports information about the current configurations. It controls the reallocation of resources to the RANs.

The RAN measurement collector (RMC) collects RAN context information and sends such information to the NRM. It can be centralized in a unique node or it can be distributed; that is an RMC instance can be active in each base station or in general in each reconfigurable node. Figure 6.9 shows RRC and RMC functionalities with interfaces.

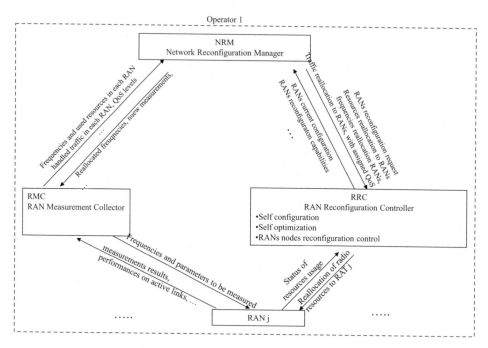

Figure 6.9 RRC and RMC functionalities with interfaces.

6.3.4 Terminal Equipment Entities

The terminal is supposed to be a multistandard device, able to follow the reconfigurations of network parameters and the changes in spectrum utilization. In the IEEE 1900.4 functional architecture, three entities are provided in the terminal equipment.

The terminal reconfiguration manager (TRM) is the entity that manages the reconfiguration of the terminal. It acquires information about the context from the terminal measurement collector (TMC) and receives from the terminal reconfiguration controller (TRC) the terminal current configuration and reconfiguration capabilities. The TRM also receives from the network reconfiguration manager (NRM) the radio resources selection policies. The output of the TRM gives inputs to the TRC and TMC of the mobile device. To the TRC, it requests the terminal reconfiguration; to the TMC, it orders new measurements. Figure 6.10 shows the entities in the terminal equipment with interfaces.

6.3.5 IEEE 1900.4 and ETSI RRS Functional Architecture Comparison

The IEEE 1900.4 functional architecture for reconfigurable radio systems has an equivalent functional architecture (FA) proposed inside ETSI and described in Chapter 5 of this book. The ETSI RRS FA is shown in Figure 6.11 and includes the following functional blocks:

• DSM (dynamic spectrum management): it manages the operator spectrum resources. It provides the bandwidth allocations to the dynamic self-organizing network planning

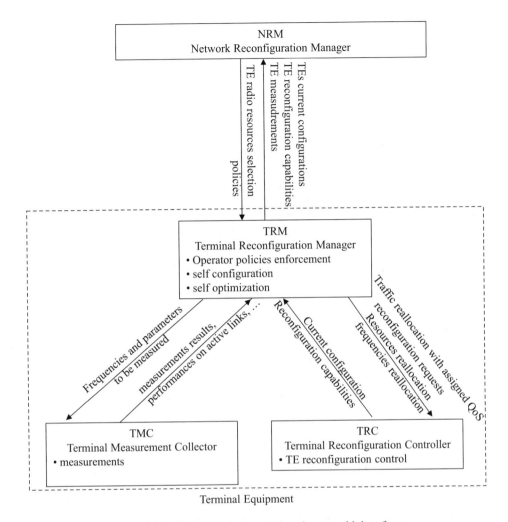

Figure 6.10   Entities in the terminal equipment with interfaces.

management (DSONPM), which maintains the context updated and gives to each RAT the available amount of spectrum. Moreover, if available, it trades the spectrum with the other operators. The DSM is aided in the decision making process by the information coming from the DSONPM entity, which provides to the DSM indicators like spectral efficiency, spatial spectrum usage, etc. Messages between DSM and DSONPM are exchanged through the MS interface.

- DSONPM (dynamic self-organizing network planning management): it plans the long term reconfiguration actions for the multi-RAT network. It receives from the configuration control module (CCM) the current configuration and mobile devices capabilities. The DSONPM receives from the JRRM information about the context and the used resources and frequencies of the mobiles. It then runs a self-optimization process leading to RAT network nodes and mobile terminal reconfiguration.

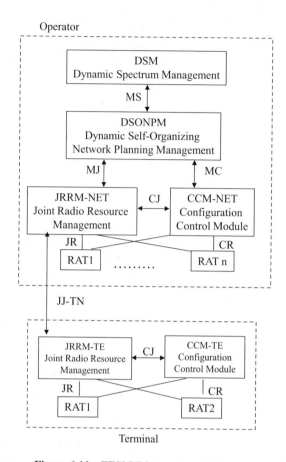

Figure 6.11 ETSI RRS standard architecture.

- CCM (configuration control module): it is implemented in the network and terminal side and controls the reconfiguration of network and terminals.
- JRRM (joint radio resource management): it is implemented in the networks and terminal side and manages the radio resources of network and terminals.

The functional blocks of the ETSI architecture are connected through standard interfaces.

Figure 6.12 shows that the functional architectures proposed inside ETSI and IEEE include different functional blocks, with some overlapping features, which lead to equivalent architectures for reconfigurable radio systems [7]. Table 6.1 shows the correspondence among the features of ETSI and IEEE functional blocks for reconfigurable radio systems [7].

OSM functionalities are included in the DSM, which also includes part of the NRM functionalities. The NRM implements DSONPM and part of the CCM and JRRM functionalities. RRC functionalities are implemented in the CCM in the network side. The RMC includes part of the CCM and of the JRRM functionalities in the network side. In the terminal side, CCM includes TRC and part of the TRM and TMC; JRRM includes part of the TMC and TRM functionalities. Table 6.1 shows the relationship between IEEE 1900.4 FA and ETSI RRS FA modules.

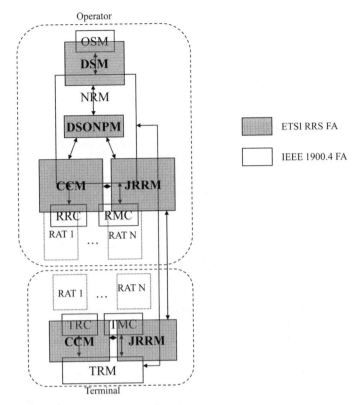

Figure 6.12   IEEE 1900.4 and ETSI RRS functional architecture comparison.

6.3.6   Use Cases for the IEEE 1900.4 Functional Architecture

The scenario considered for the proposed use cases is a heterogeneous radio access network managed from the same operator. Spectrum sharing functionalities allow dynamically distributing spectrum resources among the radio access networks (RANs) with the goal of radio resources optimization.

The entities involved in the given scenario are, in the operator network side, the radio base stations belonging to the radio access networks (RANs), the RAN measurement collector (RMC), the RAN reconfiguration controller (RRC), the network reconfiguration manager (NRM) and the operator spectrum manager (OSM). The RBSs are considered to be reconfigurable nodes equipped with hardware processing resources (HPRs) shared among different radio access technologies (RATs), like GSM/GPRS/EDGE, UMTS/HSPA/HSPA+, LTE/LTE advanced, etc. Each RAT is a radio application using a certain amount of HPRs depending of the configuration of radio parameters, the amount of ongoing traffic, etc. The HPR is connected to the radio frequency (RF) block, which goes to the antenna and is able to support different RATs at the same time.

In the terminal side, the terminal equipment (TE) includes the terminal reconfiguration controller (TRC), the terminal measurement collector (TMC) and the terminal reconfiguration manager (TRM).

Table 6.1   Relationship between IEEE 1900.4 FA and ETSI RRS FA

IEEE 1900.4 FA	ETSI RRS FA
NRM	• Part of DSM responsible for decision making on distribution of spectrum between RATs • DSONMP • Network JRRM excluding: ○ Obtaining context information from different RATs on network side • Part of network CCM responsible for execution of base station self-configuration and self-optimization

RRC	• Network CCM excluding:
	  ○ Part included into NRM
	  ○ Part responsible for obtaining context information from different RATs on network side

RMC	• Part of network CCM responsible for obtaining context information from different RATs on network side • Part of network JRRM responsible for obtaining context information from different RATs on network side
TRM	• Part of terminal CCM functionality responsible for self-configuration and self-optimization • Terminal JRRM excluding obtaining context information from different RATs in terminal

TRC	• Terminal CCM excluding:
	  ○ Part included in TRM
	  ○ Part responsible for obtaining context information from different RATs in terminal

TMC	• Part of terminal CCM responsible for obtaining context information from different RATs in terminal • Part of terminal JRRM responsible for obtaining context information from different RATs in terminal
OSM	• Part of DSM responsible for obtaining regulatory framework and for spectrum trading between operators

Reproduced with permission. Copyright European Telecommunications Standards Institute 2008. Further use, modification, copy and/or distribution are strictly prohibited. ETSI standards are available from http://pda.etsi.org/pda/

   The NRM has the task of assigning the spectrum to different RANs and evaluating possible RBS reconfigurations; the RRC jointly handles the RAN radio resources, performing radio bearer control, radio admission control, scheduling of resources in uplink and downlink, load balancing, etc. Finally, the RMC collects the measurements and provides the context to the NRM. RMC and RRC can be implemented inside the RBSs or can be external entities.

   In the described use cases, RBS, RMC and RRC are three distinct entities. The use cases in this section are proposals of the author and not of the standard. The proposed use cases are related to:

- Terminal power on and RAN selection. In this case a primary network is selected. The possible configurations are two: network driven or terminal driven. In the first case (network driven) the network is aware of the context and of the terminal capabilities and decides which radio access network the terminal will be associated with and will use for data transmission with the corresponding radio parameter configuration. In this case the network makes choices to optimize global network performances. In the second case (terminal driven) the terminal selects, depending on the context, the RAN for association and data transmission. The terminal makes choices to optimize its own resources.
- Spectrum sharing. This use case deals with an intrasystem radio resource reconfiguration, with RBSs belonging to the same operator and where the spectrum assigned to RAN1 changes according to traffic needs.
- Flexible spectrum allocation between different RATs of the same RBS.
- Radio resource optimization procedure with RBS software reconfiguration.

Figure 6.13 shows an example of the high level information flow related to the case of terminal power on and network driven RAN selection. After power on, the terminal makes radio

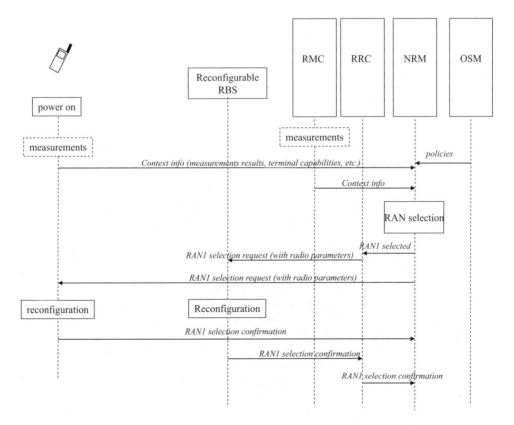

Figure 6.13 Example of high level information flow related to the case of terminal power on and network driven RAN selection.

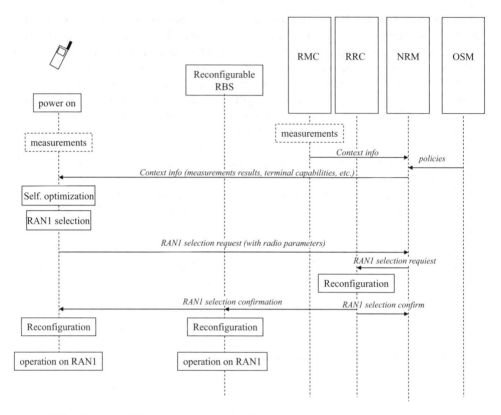

Figure 6.14 Example of high level information flow related to the case of terminal power on and terminal driven RAN selection.

measurements and sends measurement results other than its capabilities to the NRM. The NRM receives the policies from the OSM and context information from the measurement collectors in the network and terminal side. The NRM runs the radio resource optimization algorithm to choose the RAN for terminal operation that optimizes the network radio resources. This information is communicated from the NRM to the RRC, which controls the reconfiguration of RBS radio parameters for terminal operation on RAN1. At the same time, the NRM sends a request of operation on RAN1 to the terminal. After the reconfiguration of RBS and terminal radio resources, the selection of RAN1 is confirmed.

Figure 6.14 shows an example of the high level information flow related to the case of terminal power on and terminal driven RAN selection. After power on, the terminal makes is own measurements and receives the context information from the network resource manager. Then it runs a self-optimization algorithm for the selection of the radio access network that optimizes its radio parameters. It transmits the chosen RAN to the NRM to start the network reconfiguration for operation with the terminal on the selected RAN.

Figure 6.15 shows an example of the high level information flow related to the case of intrasystem radio resource reconfiguration. In the figure, RAN1 (i.e., UMTS) of RBS1 is using band B1 (i.e., two carriers of 5 MHz); RAN2 (i.e., UMTS) of RBS2 is using band

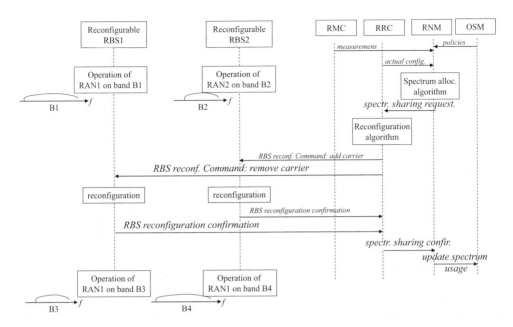

Figure 6.15 Example of high level information flow for the intrasystem radio resource reconfiguration use case.

B2 (i.e., one carrier of 5 MHz). The radio resource optimization function of the NRM entity decides that spectrum sharing between the two RBSs is needed and that one carrier must be added to RBS2 and one carrier must be removed from RBS1. RBS reconfiguration commands are sent from the RRC to RBS1 and RBS2 and the spectrum is successfully reassigned between the two base stations. After that, the OSM is updated with the spectrum sharing results.

Figure 6.16 shows an example of flexible spectrum allocation between different RANs of the same RBS. In the figure, the reconfigurable radio base station (RBS) is simultaneously running RAN1 (i.e., UMTS) in bandwidth B1 and RAN2 in bandwidth B2 (i.e., GSM). The spectrum allocation function of the NRM decides to reallocate the spectrum between RAN1 and RAN2. The NRM communicates the spectrum reallocation to the RRC. The RRC reconfiguration algorithm decides the radio parameters for operation of RAN1 and RAN2 in the new bands. An RBS reconfiguration command is sent to RBS1 and the spectrum is successfully reassigned between the two radio access technologies. The NRM and OSM are updated with the successful spectrum reassignment.

The last proposed use case is an example of the radio resource optimization procedure. The high level information flow is shown in Figure 6.17. During the ongoing phase, the terminal uses RAN2 for data transmission. The NRM receives the policies from OSM and context information from the measurement collectors in the network and terminal sides. The NRM runs the radio resource optimization algorithm, which gives RAN3 as preferred. This information is communicated from the NRM to the RRC, which controls the reconfiguration of RBS radio parameters from RAN2 to RAN3 operation. At the same time, the NRM sends a request of RAN3 selection to the terminal. After the reconfiguration of RBS and terminal radio resources, the selection of RAN3 is confirmed.

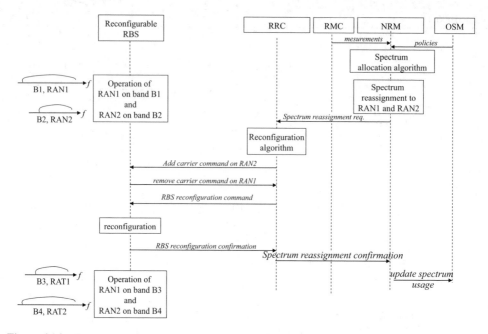

Figure 6.16 Example of high level information flow for flexible spectrum allocation between different RANs of the same RBS.

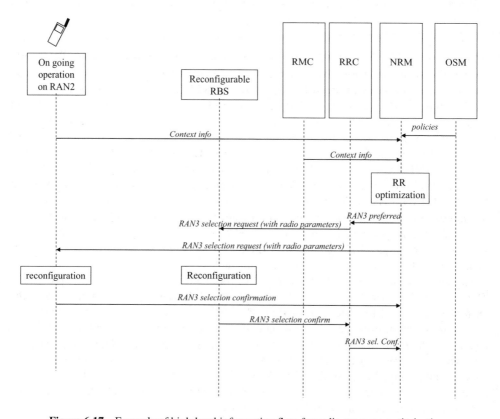

Figure 6.17 Example of high level information flow for radio resource optimization.

6.4   IEEE 1900.4a Functional Architecture

In 2011, an amendment to the IEEE 1900.4a standard came out, which defined the architecture and the interfaces for dynamic spectrum access networks in white space frequency bands [8]. White spaces refer to the portion of available spectrum in the VHF and UHF bands left from the transition of television channels from analogue to digital. The efficient use of white spaces is very important for an optimal utilization of this valuable part of the radio spectrum. The IEEE 1900.4a introduces to the IEEE 1900.4 functional architecture some blocks enabling white space transmission without any limitation on the used radio interfaces. The IEEE 1900.4a functional architecture (FA) is presented in Figure 6.18.

In Figure 6.18, the IEEE 1900.4 functional architecture is added to a white space (WS) radio access network (RAN), with the following functional blocks:

- Cognitive base station (CBS): it is a base station capable of using white spaces without interfering with the primary users (i.e., existing RANs).
- CBS reconfiguration manager (CBSRM): it manages the reconfiguration of the CBS. It is aided in the optimization process from the context information provided by the CBS and the information coming from the cognitive terminals. In particular, the CBSRM and TRMs exchange information about the classification of white space resources in order to allow the cognitive network to exist as a secondary network without interfering with the primary users. The connection of the CBSRM with the network reconfiguration manager (NRM) of the legacy radio access networks guarantees the optimal usage of the white space spectrum with the joint optimization of the radio resources of licensed RANs. The CBSRM

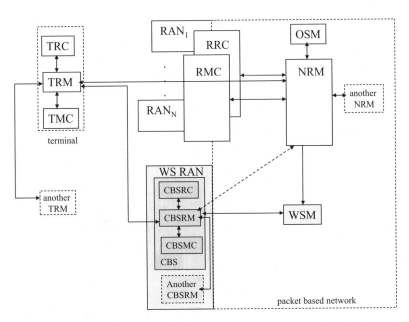

Figure 6.18   High level IEEE 1900.4a functional architecture and the interfaces for dynamic spectrum access networks in white space frequency bands.

receives from the white space manager (WSM) the regulatory context and transmits the used frequencies with relative powers for white space database updating. It communicates with other CBSRMs for coordination of white space.

• CBS measurement collector (CBSMC): it collects the CBS context information and sends it to the CBSRM.

• CBS reconfiguration controller (CBSRC): it receives reconfiguration requests from the CBSRM and controls the radio access network reconfiguration.

• White space manager (WSM): it provides the regulatory context, like the available frequencies for secondary usage and relative allowed powers, etc. It acts as a white space database. It enables the collaboration between IEEE 1900.4a and IEEE 1900.4 systems.

In the terminal side, the functional blocks of the IEEE 1900.4a architecture are the same as the IEEE 1900.4. The central functional block in the terminal side is the terminal reconfiguration manager (TRM), which communicates with the CBSRM in the network side in order to optimize the spectrum usage radio parameters in the white space frequency bands. The TRMs of cognitive radio devices can directly communicate to exchange measurements, spectrum sensing results and context information (i.e., geo-location information, terminal capabilities, etc.). The CBSRMs of a white space network can also communicate with other CBSRMs in order to coordinate the usage of white spaces. They enforce the policies received from the white space manager and communicate these to the spectrum used and the related radio parameters (i.e., maximum transmitted power). Figure 6.19 shows an example of information exchange among the functional blocks added in the IEEE 1900.4 architecture.

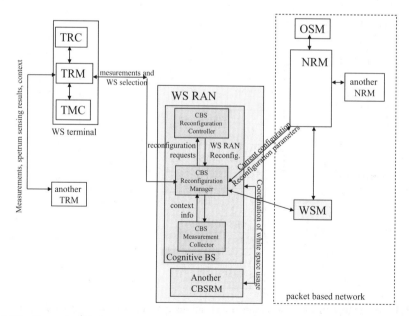

Figure 6.19   Example of information exchange among functional blocks of IEEE 1900.4 architecture.

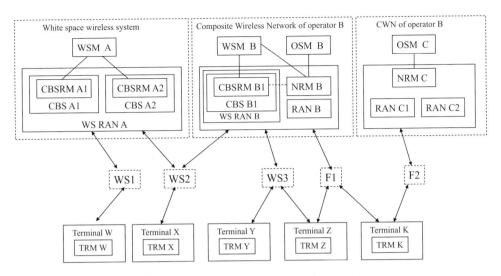

Figure 6.20 IEEE 1900.4a scenario example.

The IEEE 1900.4a functional architecture enables a scenario where a reconfigurable radio device is able to simultaneously communicate with a white space secondary network and with a primary network. This scenario is presented in Figure 6.20.

In the example of Figure 6.20, three networks are represented: the white space (WS) wireless system A; the composite wireless network (CWN) of operator B including a WS wireless system and radio access networks (RANs) managed by a network resource manager (NRM); the CWN of legacy RANs of operator C. In this scenario, terminal W operates in the WS frequency WS1 through the WS RAN of operator A; terminal X operates in the WS frequency WS2 through the WS RANs of both operators A and B; terminal Y operates in the WS frequency WS3 through the WS RAN of operator B; terminal Z operates in the WS frequency WS3 through the WS RAN of operator B and in the frequency F1 through RANB of the same operator (B); terminal K operates in F1 of RANB with operator B and in F2 of RANC1 with operator C. Then, the IEEE1900.4a functional architecture enables spectrum sharing with other 1900.4a functional architectures, with IEEE 1900.4 systems and also with non-IEEE 1900.4 systems (case not represented in the figure) [8].

6.4.1 White Space Manager Entity

The white space manager (WSM) is the entity that provides regulatory white space policies to the cognitive base station reconfiguration manager (CBSRM) and acts as a white space database. It also enables collaboration between IEEE 1900.4 and IEEE 1900.4a systems. The WSM communicates with the CBS reconfiguration manager (CBSRM) to add a CBS to the WS database at the end of the registration procedure; to receive context and location information from the CBSs, other than the queries for getting the list of neighbouring CBSs, the state of a particular CBS, etc. The WSM deletes a CBS from the WS database after its deregistration. Figure 6.21 shows the WSM functionalities with interfaces.

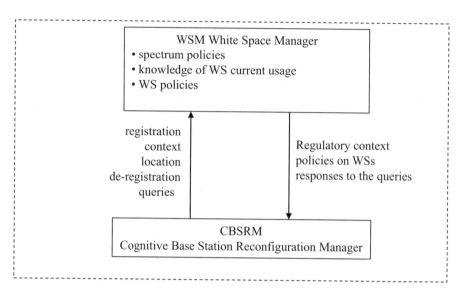

Figure 6.21 WSM functionalities with interfaces.

6.4.2 *Cognitive Base Station*

The cognitive base station (CBS) in the IEEE 1900.4a architecture is an access network node able to sense the environment and flexibly use the white spaces of the radio spectrum. The cognitive base station is capable of self-reconfiguration and self-optimization. It includes three functional entities: the cognitive base station reconfiguration manager (CBSRM), the cognitive base station reconfiguration controller (CBSRC) and the cognitive base station measurement collector (CBSMC).

6.4.2.1 Cognitive Base Station Entities

The cognitive base station reconfiguration manager (CBSRM) manages the white space (WS) resource usage. It classifies WSs, enforces WS policies received from the WSM, is provided by self-configuration and self-optimization algorithms and manages spectrum sensing measurements. It communicates with the terminal reconfiguration managers (TRMs) and coordinates with other CBSRMs and NRMs in order to assure the coexistence among primary and secondary systems. The CBSRM obtains context information from the CBS measurement collector (CBSMC) and is able to request to it to perform specific measurements. In the communication with other TRMs, the CBSRM sends other related context information to other CBSs and eventually to other terminals. It also provides the TRM radio resource selection policies. The TRMs send to the CBSRM the WS classification, the measurements, the current configuration and reconfiguration capabilities.

The CBSRM exchanges with other CBSRMs CBS context information, terminals context information, WS database information, reconfiguration decisions, etc. It is also connected to

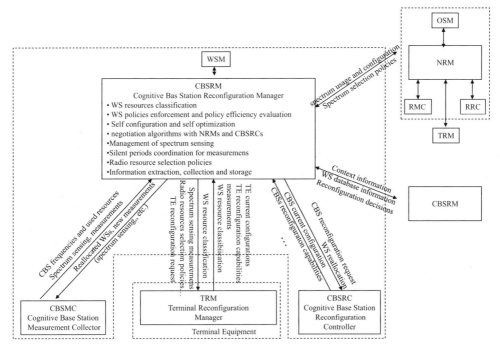

Figure 6.22   CBSRM functionalities with interfaces.

the CBS reconfiguration controller to send CBS reconfiguration requests and to assign WSs. Figure 6.22 shows the CBSRM functionalities with interfaces.

6.4.3   Terminal Equipment Entities

The terminal is supposed to be a multistandard device, able to dynamically change the used spectrum and radio parameters. In the IEEE 1900.4a functional architecture (FA), as in the IEEE 1900.4 FA, three entities are provided in the terminal equipment: the terminal reconfiguration manager (TRM), the terminal measurement collector (TMC) and the terminal reconfiguration controller (TRC).

The terminal reconfiguration manager (TRM) is the entity that manages the reconfiguration of the terminal. It acquires information about the context from the terminal measurement collector (TMC) and receives from the terminal reconfiguration controller (TRC) the terminal current configuration and reconfiguration capabilities. The TRM also receives from the cognitive base station reconfiguration manager (CBSRM) the radio resources selection policies. In the IEEE 1900.4a architecture, the TRM manages in the terminal side the WS resources, enforcing the policies received from the CBSRM.

The output of the TRM gives inputs to the TRC and TMC of the mobile device. To the TRC, it requests the terminal reconfiguration including WS reallocation; to the TMC, it orders

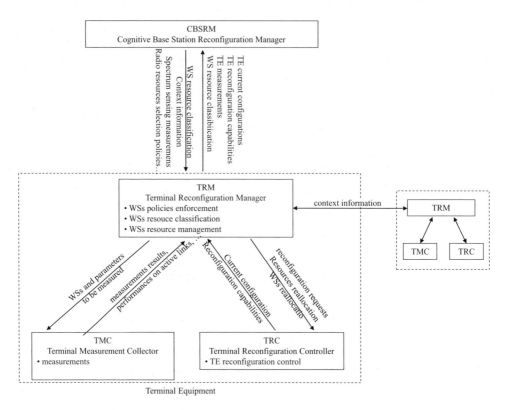

Figure 6.23   Entities in the terminal equipment with interfaces.

new measurements. Moreover, TRMs of different terminals can exchange context information. Figure 6.23 shows the entities in the terminal equipment with interfaces.

6.4.4   Use Cases for the IEEE 1900.4a Functional Architecture

Several examples and use cases can be designed for the IEEE 1900.4a functional architecture. In this section the use cases are proposed as high level information flow among the functional nodes. The use cases here are proposals of the author and not of the standard.

The considered scenario is of a heterogeneous radio access network, with primary or secondary usage of the spectrum, belonging to the same or different operators, with reconfigurable radio capabilities. The proposed use cases are related to:

- Terminal power on and WS selection. This use case is related to the selection of secondary frequency bands for the establishment of a communication.
- White space operation flexibility. This use case shows the flexible white spaces allocation.

Figure 6.24 shows an example of information flow for power on and WS selection. After power on, the terminal measurement collector (TMC) senses the WS band for measurements

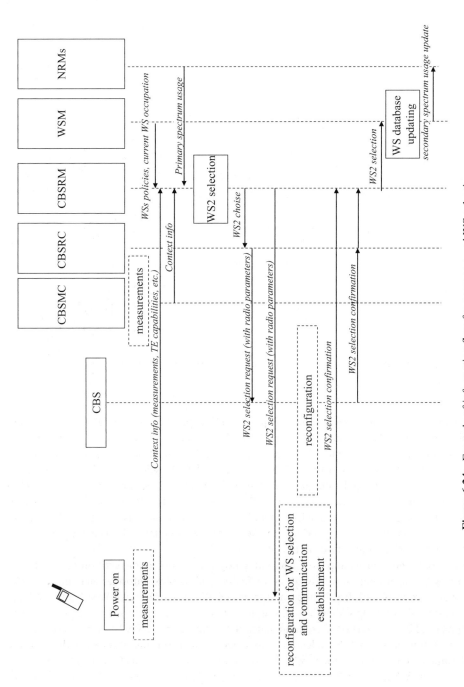

Figure 6.24 Example of information flow for power on and WS selection.

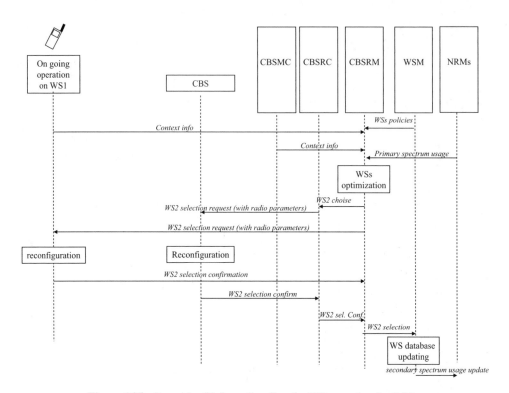

Figure 6.25   Example of information flow for WS operation flexibility.

and sends the results and the terminal equipment (TE) capabilities to the CBSRM, which also receives the CBSMC measurements. The CBSRM is able to complete context information because it receives the current WS usage other than WS occupation from the WSM, and the primary spectrum usage from the NRMs. Then the CBSRM selects a WS (in the example WS2) for TE operation and communicates the choice to the CBSRC, which controls the CBS reconfiguration for WS2 selection. The CBSRM also controls, together with the TRM, the terminal reconfiguration for WS selection and communication establishment. Finally, the WS selection is confirmed to the WSM and the WS database is updated. If necessary, the NRMs of the IEEE 1900.4 architecture are also updated.

Figure 6.25 shows an example of information flow for WS operation flexibility. In the figure, the terminal is operating using WS1. It periodically sends context information to the CBSRM, which also receives the CBSMC measurements and WS usage from the WSM. The CBSRM runs an optimization algorithm, whose output is frequency change to WS2 for the terminal equipment (TE). The CBSRM communicates the output to the CBSRC, which controls the CBS reconfiguration for WS2 selection. The CBSRM also controls, together with the TRM, the terminal reconfiguration for WS2 selection. Finally, the WS2 selection is confirmed to the WSM and the WS database is updated. If necessary, the NRMs of the IEEE 1900.4 architecture are also updated.

6.5 Summary

In 2009, the IEEE published the IEEE 1900.4 standard, which proposes a functional architecture enabling network device distributed decision making for optimized radio resource usage in heterogeneous wireless access networks [1]. The proposed functional architecture envisages scenarios with a single operator and different network operators collaborating for the joint goal of radio resource optimization.

New entities are introduced for operator spectrum management, network and terminals reconfiguration management, reconfiguration control and measurement collection. The new entities bidirectionally communicate through standard interfaces and with their corresponding entity in the radio terminal.

In 2011, an amendment to the IEEE 1900.4 standard, the IEEE 1900.4a, was published. IEEE 1900.4a adds to the IEEE 1900.4 functional architecture new entities for dynamic spectrum access networks in white space frequency bands. The new entities add cognitive features to cognitive base stations (CBSs), such as CBS reconfiguration management, CBS reconfiguration control and CBS measurement collection, and look at the possibility of enabling secondary transmission using white spaces.

References

1. IEEE Standard 1900.4 (2009) *Architectural Building Blocks Enabling Network-Device Distributed Decision Making for Optimized Radio Resource Usage in Heterogeneous Wireless Access Networks*, 27 February 2009.
2. Guenin, J. (2008) IEEE SCC41 Standards for Dynamic Spectrum Access Networks, Joint Workshop with SCC41, SDRF and IEICE SRTC, August 2008.
3. http://grouper.ieee.org/groups/dyspan/index.html.
4. Murroni, M., Prasad, R.V., Marques, P. *et al.* (2011) IEEE 1900.6: Spectrum Sensing Interfaces and Data Structures for Dynamic Spectrum Access and Other Advanced Radio Communication Systems Standard: Technical Aspects and Future Outlook. *IEEE Communications Magazine*, **49** (2), December.
5. Buljore, S., Harada, H., Filin, S. *et al.* (2009) Architecture and enablers for optimized radio resource usage in heterogeneous wireless access networks: the IEEE 1900.4 working group. *IEEE Communications Magazine*, **47** (1), January, 122–129.
6. Prasad, R.V., Pawlczak, P., Hoffmeyer, J.A. and Berger, H.S. (2008) Cognitive functionality in next generation wireless networks: standardization efforts. *IEEE Communications Magazine*, **46** (4), April, 72–78.
7. ETSI TR 102 682 (2009) *Reconfigurable Radio Systems (RRS); Functional Architecture for Management and Control of Reconfigurable Radio Systems*, v1.1.1.
8. IEEE Standard 1900.4a (2011) *Architectural Building Blocks Enabling Network-Device Distributed Decision Making for Optimized Radio Resource Usage in Heterogeneous Wireless Access Networks – Amendment 1: Architecture and Interfaces for Dynamic Spectrum Access Networks in White Space Frequency Bands*.

7

Regulatory Challenges of Reconfigurable Radio Systems

7.1 Introduction

As stated in the first paragraph of this book, the evolution of the network in the recent past regarded the introduction of new access technologies, both fixed and wireless, using an IP backbone for all originating and terminating services (see Figure 7.1).

Today, licensed and unlicensed systems offer to the end user the possibility to be connected through various frequency bands and with different speeds and provided services. Emerging technologies like cognitive [1] and reconfigurable radio devices and network nodes will enable new network scenarios and new modes of spectrum usage. This will be facilitated by the development of new spectrum policies that make the use of new technologies effective [2, 3].

This chapter, without pretending to be exhaustive, presents the evolution of spectrum management in the case of a shared access among different entities, with corresponding scenarios.

7.2 Spectrum Management

Spectrum management meets the spectrum requirements of fixed and mobile, terrestrial and satellite services, governmental and military networks, broadcasters and other users [4]. The growing demand of wireless services and the need for higher and higher speeds for data connection lead to a very high demand of the scarce resource which is the spectrum.

The spectrum is managed at a global level from the International Telecommunication Unit (ITU). The key documents are the International Telecommunication Union (ITU) Radio Regulations (RRs), which are recognized as international treaties. Radio Regulations set the context for the National Regulatory Authorities to license a radio spectrum, containing general rules for the assignment and use of frequencies and including a Table of Frequency Allocations of the various radio services. The spectrum is also managed at regional and national levels, addressing political, regulatory and technical aspects. ITU divides the world into three regions, represented in Figure 7.2.

Reconfigurable Radio Systems: Network Architectures and Standards, First Edition. Maria Stella Iacobucci.
© 2013 John Wiley & Sons, Ltd. Published 2013 by John Wiley & Sons, Ltd.

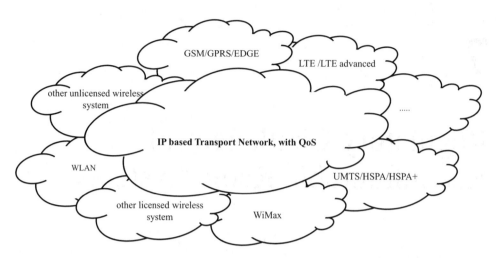

Figure 7.1   Wireless access technologies connected to an IP backbone.

Almost all countries/territories fall within a Regional Regulatory Group:

- CEPT/ECC: the Electronic Communication Committee (ECC) inside CEPT (European Conference of Postal and Telecommunications Administration) 'considers and develops policies on electronic communications activities in European context, taking account of European and international legislations and regulations' [5].
- CITEL: the Inter-American Telecommunication Commission (CITEL), entity of the Organization of American States (OAS), is a 'region's premier inter-governmental telecommunication advisory body' [6].
- APT: the Asia Pacific Telecommunity (APT) serves as the focal organization for ICT in the Asia Pacific region [7].
- Arab States.
- Africa.

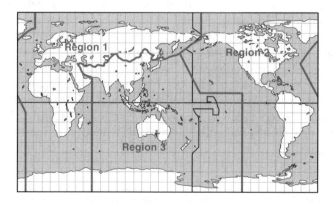

Figure 7.2   ITU regions.

ITU Radio Regulation article 5 states that each band can be allocated to different services and that each National Regulatory Authority can choose the service to license in the band, with the goal to minimize harmful interference.

Up to now, the usage of spectrum from telecommunication radio systems can be divided into:

- Licensed usage of spectrum: for licensed systems the spectrum usage is exclusive. In this case, one entity uses the licensed portion of the spectrum.
- Unlicensed usage of spectrum: unlicensed systems are allowed to transmit on an unlicensed band with given parameters, like maximum power. For example, in the unlicensed band from 2.4 to 2.4835 GHz different systems operate, like WLANs (IEEE 802.11), Bluetooth (IEEE 802.15.1), RFiD (ISO 18092), etc. All these systems can use a maximum power of 100 mW e.i.r.p. There is no limitation on the number of systems and devices on such unlicensed bands and the risk is that everyone experiences bad performances. This is why unlicensed wireless systems are not able to guarantee quality of service (QoS).

Figure 7.3 shows the usage of licensed and unlicensed bands.

In managing the spectrum, most regulators use a command and control (CAC) approach. CAC can be defined as 'the direct regulation of an industry or activity by legislation that states what is permitted and what is illegal' [8]. This means that it is regulated by how each part of the radio spectrum can be used and the entity that can use it. For example, the Federal Communication Commission (FCC) uses a CAC approach based on the following five steps [9]:

- Allocation: the spectrum fragments (bands) are allocated to different types of services.
- Adoption of service rules or technical standards: service rules and technical standards are adopted in each allocated band. Technical and operational characteristics (i.e., power limits, carrier spacing, etc.) of the radios that will use the allocated spectrum are established.

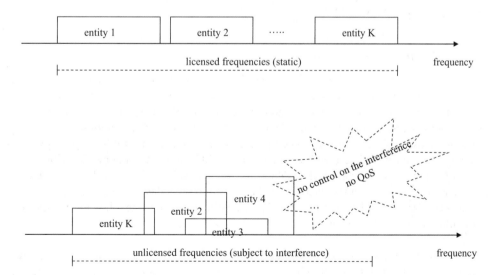

Figure 7.3   Usage of licensed and unlicensed bands.

Figure 7.4 Five steps of the CAC approach in spectrum management.

- Certification: major systems must be certified.
- Assignment: the different portions of the spectrum are assigned to technical standards or service rules.
- Enforcement: the assignments are enforced through measurements, reporting, etc.

Figure 7.4 shows the five steps of the CAC approach in spectrum management.

The CAC approach leads in the United States to the 'FCC table of frequency allocations' [10, 11] and in Europe to the 'European table of frequency allocations and applications in the frequency range 9 kHz to 3000 GHz' (ECA Table) [12].

7.2.1 Dynamic Spectrum Access

Dynamic spectrum access (DSA) introduces flexibility in spectrum management with the twofold goal of allowing new users to transmit on the available frequencies with constraints (i.e., maximum power, etc.), while maintaining control on the interference. Such flexibility brings political, regulatory and technical challenges for all the involved entities [13].

Licensed users obtain (with the license) a primary right of spectrum usage. However, licensed spectrum is underutilized. As stated from FCC's Spectrum Policy Task Force (ET Docket 02-135), Ofcom's Spectrum Framework Review and the European Unit Framework Directive, some of the licensed bands could be opened to secondary spectrum usage.

The secondary use of the spectrum [14, 15] can be categorized into the following approaches, with different levels of constraint:

- The secondary users must only avoid harmful interference to primary users. In this case, the secondary users operate on an unlicensed basis, with no constraints on the technology or applications. Quality of service (QoS) cannot be guaranteed to the secondary users.

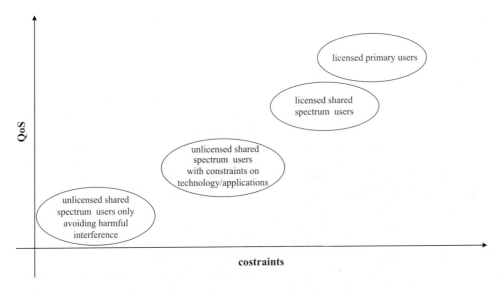

Figure 7.5   Different spectrum usages related to constraints and QoS.

- The secondary users operate on an unlicensed basis, but have constraints on the technology or applications. In this case QoS cannot be guaranteed to secondary users.
- The secondary users operate on a licensed basis with defined sharing rules. In this case QoS can also be guaranteed to secondary users.

Figure 7.5 shows the different spectrum usages related to constraints and QoS possibilities.

The dynamic spectrum access model involves the transition from an exclusive usage of spectrum regulated from individual licenses to dynamic spectrum management. This evolution is represented in Figure 7.6 towards two steps: unlicensed shared spectrum and licensed shared access (LSA).

The unlicensed shared spectrum includes all the cases where a secondary user transmits without license, avoiding interference to primary users, eventually with constraints. In the simplest scenario, devices try to avoid interference without explicit signalling. They are able to sense the presence of other transmitters. Secondary devices that sense the presence of

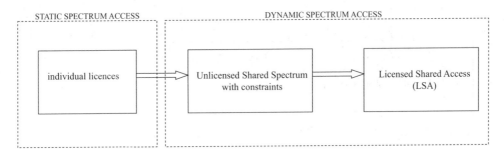

Figure 7.6   Transition to dynamic spectrum access.

secondary systems interrupt their transmission and move to another frequency, eventually rearranging radio parameters.

Licensed shared access is the case where secondary users operate with licenses and defined sharing rules. LSA is defined in [16] as 'an individual licensed regime of a limited number of licensees in a frequency band, already allocated to one or more incumbent users, for which the additional users are allowed to use the spectrum (or part of the spectrum) in accordance with sharing rules included in the rights of use of spectrum granted to the licensees, thereby allowing all the licensees to provide a certain level of QoS'.

The sharing agreements can include quality of service guarantees for the secondary systems and also a secondary spectrum market (i.e., the possibility of spectrum renting). In this case, new markets could be foreseen with spectrum offering and spectrum renting on a time and frequency basis. A certain band could be used for short or long periods depending on the needs, availability and other parameters.

Recently, an industry consortium introduced the term authorized shared access (ASA) to provide shared access to an IMT spectrum under a licensing regime in order to offer services with a certain quality of service. ASA and LSA can be considered synonyms.

Figure 7.7 summarizes the possible spectrum usages described in this section. The first distinction is between licensed and unlicensed usage of the spectrum. Unlicensed spectrum usage is also known as 'the commons', in which everyone shares a common band, eventually with some constraints, without paying any fee for spectrum usage. In this scenario the only protection from interference is given by the constraints, like a maximum allowed power, and robust techniques at physical and MAC layers. An example of commons is IEEE 802.11, whose transmission in the bandwidth of 2.4 GHz is limited at 100 mW e.i.r.p. and uses a spread spectrum or OFDM at the physical layer and CSMA/CA at the MAC layer. In the same band also other systems like Bluetooth, RFid, etc., are allowed to transmit.

In a scenario of secondary spectrum usage, licensees acquire the primary usage of the spectrum, while secondary usage of the spectrum can be obtained with or without licenses. In this scenario, collective use of spectrum (CUS) [16–19] is defined in [16] as follows:

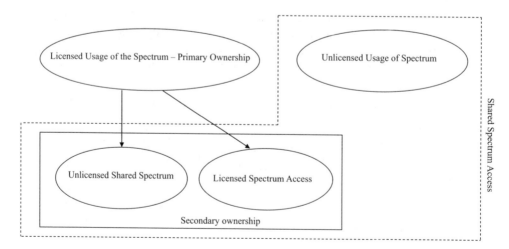

Figure 7.7 Possible spectrum usages.

'Collective Use of Spectrum allows an unlimited number of independent users and/or devices to access spectrum in the same range of designated CUS frequencies at the same time and in a particular geographic area under a well-defined set of conditions.'

Three entities are responsible for CUS [16]:

• The regulator, which sets the conditions,
• The manufacturer, which ensures that the equipment conforms to the specified set of conditions, and
• The user, which uses a conform terminal.

The CUS model does not guarantee quality of service (QoS), which can be assured only by the licensed shared access (LSA) model.

Under the CUS model, two different implementations can be taken into account [16]:

• Light licensing model: the regulator limits the number of users accessing the bandwidth. In this case, higher power levels than those typically employed for licence-exempt applications can be used.
• Private commons: the rules that determine access to the band are set by the entity to which the band has been licensed.

Figure 7.8 shows the described possible implementations of CUS.

Advanced forms of spectrum sharing are cooperative systems [20]. In cooperative systems, all the terminals and network nodes cooperate to assure an optimized usage of the spectrum. The simplest case of cooperative systems is when two network elements find an agreement on

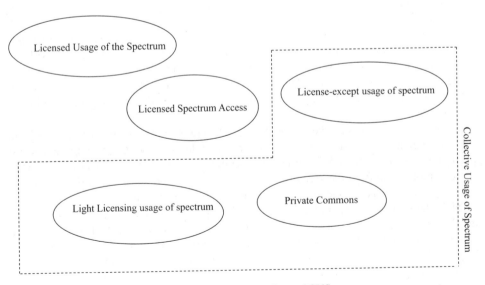

Figure 7.8   Implementations of CUS.

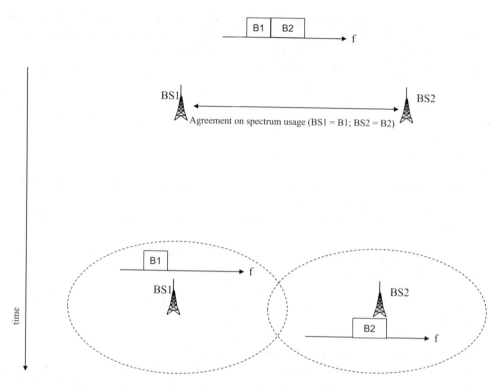

Figure 7.9   Example of a simple case of coordinated spectrum sharing.

the band for exclusive usage. An example of this simple case of coordinated spectrum sharing is shown in Figure 7.9.

In the example, the base stations (BSs) BS1 and BS2 exchange signalling messages to agree on the operating band. After that, each base station uses the agreed band (BS1 uses B1 and BS2 uses B2) for coverage and transmission with the associated users. Cooperation can be among systems owned by the same or different subjects. If the owners of cooperative systems are different subjects, then new markets including spectrum renting and offering can be considered.

Shared spectrum access can be operated in the cooperation mode, where secondary systems cooperate with primary systems and agree on the possible usage of the spectrum. In this case, spectrum management could be distributed or centralized in a spectrum manager (SM), which receives information from the systems and coordinates the secondary usage of the spectrum. Figure 7.10 shows an example of centralized spectrum management.

The highest level of cooperation is with mobile ad hoc networks (MANETs), where each mobile terminal is a network node able to route other's information and to adapt all the parameters to optimize the whole network. In such systems spectrum management must be distributed and the same spectrum management protocols must be implemented in all the network nodes. Signalling must be used to share information. A MANET with distributed spectrum management is represented in Figure 7.11.

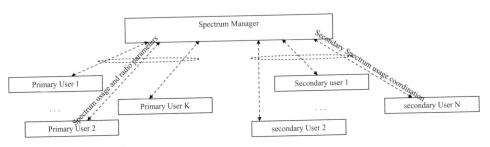

Figure 7.10 Example of centralized spectrum management.

With cognitive and reconfigurable radio technologies, the possibility of a QoS guarantee in secondary spectrum usage will be highly improved. Reconfigurable radio network nodes will be able to sense the environment, receive spectrum policies, choose the right transmission band and adapt radio parameters to the actual context (i.e., used frequencies, operating networks, etc.) in order to meet quality of service (QoS) requirements.

7.2.2   Market-Based Approach in Spectrum Management

Up to now, an administrative approach has governed the usage of spectrum. A command and control approach is the one currently employed by most regulators around the globe and is based on the two following fundamental steps:

- Spectrum allocation: the spectrum allocation is planned and technical conditions for radio spectrum usage are established.
- Spectrum assignment.

Recently, connected to the dynamic shared access principles, a market-based approach in spectrum management has emerged [21–24]. Market mechanisms represent an alternative to

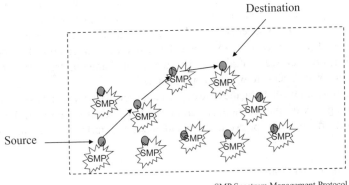

SMP Spectrum Management Protocol

Figure 7.11 Example of a MANET with distributed spectrum management.

SERVICE NEUTRALITY	SPECTRUM AUCTION
SECONDARY SPECTRUM TRADING	ADMINISTRATIVE INCENTIVE PRICING

Figure 7.12   Market-based approaches in spectrum management.

the CAC approach, putting the resource of the spectrum in the hands of the market. However, very probably the decision of whether, how and when to deploy such market-based mechanisms will remain with governments and regulation authorities. It is possible that in the future some frequencies will be left to the market and other frequencies, like the frequencies used by public emergency services, will continue to be regulated from administrative decisions.

A market-based approach in spectrum management (see Figure 7.12) includes the following main strategies [23]:

- Change of use or service neutrality: the spectrum is not connected to a technology or a service. Different services and technologies compete for the same spectrum usage.
- Auctions: typically used for primary spectrum assignment, these are based on bids. The spectrum is assigned to the firm who bids most.
- Spectrum trading: typically referred to secondary markets, it allows changes of ownership and reconfigurations based on spectrum transfers and spectrum leasing.
- Administrative incentive pricing (AIP): this method influences spectrum assignment and spectrum allocation through the application of an opportunity cost.

7.2.2.1   Service Neutrality

The spectrum is a scarce resource. Because of that, the service neutrality paradigm, also called change of use, opens to a scenario where the owner of a frequency band is allowed to use whatever technology for whatever service. This means that the use of the spectrum becomes technology neutral (i.e., the owner can change the technology, for example by switching from GSM to LTE) and/or service neutral (i.e., the owner can change the service, for example by switching from TV broadcast to wireless broadband).

Some exceptions could occur in the case of particular services, like safety services. The challenge in defining service and technology neutral rights is the introduction of flexibility while maintaining the efficiency in spectrum usage and avoiding interference to other users [25].

7.2.2.2   Spectrum Auctions

A spectrum auction is a process used to sell the licenses for spectrum usages. In general, resources are assigned to the subjects that value them the most.

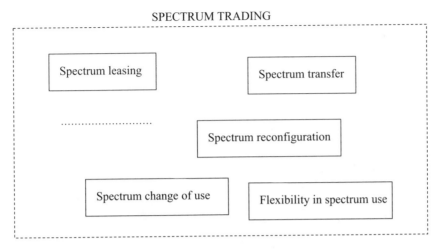

Figure 7.13   Spectrum trading.

7.2.2.3   Spectrum Trading

Spectrum trading (see Figure 7.13) is a market-based mechanism that allows secondary usage of a radio spectrum, which is secondary trading. Secondary trading permits the purchaser to change the use to which the spectrum was initially licensed while maintaining the right to use it. It includes transfer, leasing of licenses or of part of them, spectrum reconfiguration, change of use, flexibility in spectrum use, etc.

Secondary trading is seen as the first step in the reform of spectrum regulation. It opens up a more efficient usage of the spectrum and leads the way to new technologies like cognitive and reconfigurable radio systems.

7.2.2.4   Administrative Incentive Pricing

Administrative incentive pricing (AIP) is a spectrum management method that influences spectrum assignment and spectrum allocation through the adoption of a methodology for the calculation of prices that generate incentives for licensees to use the spectrum in an efficient way [25]. Spectrum prices should be set at the point of balance of cost/benefits. Figure 7.14 shows the point of balance in the case of two sectors, which can be thought of as two cells of radio access systems.

Figure 7.14 shows the marginal benefits (MBs) for sectors 1 and 2. The MBs decrease when the amount of spectrum (S) increases. The optimum is reached at the balance point (S*, MB*), which determines the optimal amount of spectrum to be assigned to sectors 1 and 2.

Economics states that the marginal benefit decreases as consumption increases. In general, the curves of demand and marginal benefits have similar shapes (decreasing as the quantity increases). In fact the law of demand is based upon the idea of diminishing marginal benefit. Because the price (and the marginal benefit) decreases when demand increases, price also decreases as consumption increases (law of demand). The balance point (S*, MB*) determines a price (fee) for the two sectors that takes into account the described low. Coming back to

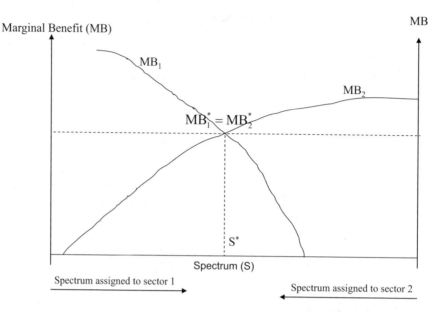

MB$_1$: Marginal Benefit of sector 1

MB$_2$: Marginal Benefit of sector 2

Figure 7.14 Equilibrium point for competing uses of the spectrum. (Reproduced with permission from "An Economic Study to Review Spectrum Pricing," Indepen, Aegis Systems and Warwick Business School, February 2004.)

Figure 7.14, if curves change, the two sectors will lose the equilibrium point. In general, the equilibrium point of optimal spectrum allocation changes if the spectrum demands (and then MBs) change [26].

Figure 7.15 shows that in the previous equilibrium point * the marginal benefit of the spectrum in sector 1 is greater than that in sector 2. Then sector 2 should have an incentive in the price it pays to return the spectrum in excess rather than pay the fee. Administrative incentive pricing is based on the principle of setting spectrum prices that achieve an optimum spectrum usage and then providing incentives for spectrum reallocations if needed. Figure 7.16 continues the different licensing regimes with spectrum management approaches.

7.3 Impacts of Reconfigurable Radio Systems to Spectrum Governance

The traditional command and control (CAC) spectrum management model presents some limitations: some frequencies are underutilized and the method slowly responds to changes in spectrum demands and new services/technologies. Then, the need for a more adaptable and flexible spectrum management has been recognized from the spectrum management authorities [27]. Such a flexible spectrum management model is also called dynamic spectrum access (DSA) or opportunistic spectrum access (OSA).

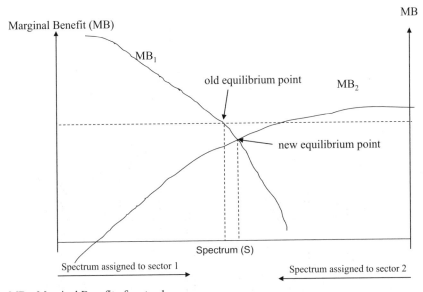

MB$_1$: Marginal Benefit of sector 1
MB$_2$: Marginal Benefit of sector 2

Figure 7.15   Competing uses of the spectrum: change of the equilibrium point.

The technologies of cognitive and reconfigurable radio systems are open to new spectrum management models. The systems described in other sections of this book – the cognitive systems IEEE 802.22, IEEE 1900.4 and ETSI standards for reconfigurable radio systems – envisage scenarios of different spectrum usages, both licensed and unlicensed, with dynamic spectrum management. Such scenarios have been described in Chapters 4, 5 and 6 of this book and are resumed in this section in relation to the dynamic spectrum management approaches described in the previous paragraph of this chapter (Section 7.2).

IEEE 802.22, the first standard based on the cognitive radio, introduces the possibility of an unlicensed transmission on the white spaces (WSs), which are the spectrum holes left in the UHF and VHF bands from the transition to digital television. Because spectrum holes depend on the location, spectrum management in IEEE 802.22 is aided by a geo-location database.

The IEEE 802.22 network topology is point-to-multipoint, with IEEE 802.22 base stations connected to the Internet backbone. The base station (BS) realizes the radio coverage and provides connectivity service to up to 512 fixed or portable stations. Each BS is provided with geo-location capabilities and communicates with the geo-location database. The database service transmits to the BSs the spectrum policies and the available TV channels, and the BSs update the database service with the used TV channels. Each BS is provided with a spectrum manager (SM), which handles channel selection and management, schedules spectrum sensing operation, enforces spectrum policies, accesses database service and protects the incumbents. IEEE 802.22 BSs are limited in power and maximum antenna eight. If the database service does not exist, the spectrum manager chooses an available channel on the basis of spectrum sensing results.

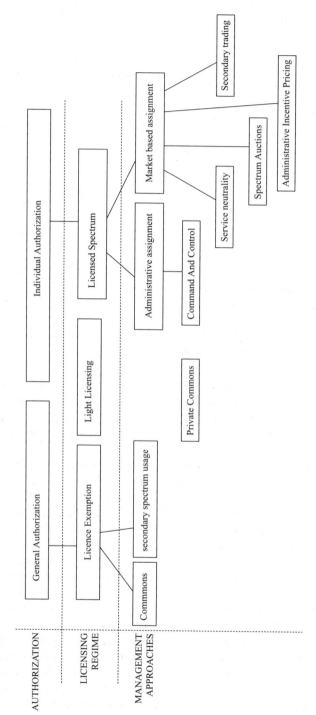

Figure 7.16 Different licensing regimes with spectrum management approaches.

Table 7.1 Spectrum management approaches of the IEEE 802.22 system

SYSTEM	Mode of operation	Licensing regime	Management approach
IEEE 802.22	Normal	License exemption	Secondary spectrum use
	Coexistence	License exemption	Shared secondary spectrum use

The BSs can operate in two modes:

- Normal mode: each BS transmits on one TV channel.
- Coexistence mode: many IEEE 802.22 BSs share the same TV channel.

Table 7.1 gives spectrum management approaches of the IEEE 802.22 system.

The ETSI and IEEE functional architectures for reconfigurable radio systems (RRSs), described in this book, present new spectrum management models, in licensed and unlicensed regimes. The scenario of RRSs is with heterogeneous access networks connected to a unique packet-based core network. Both ETSI and IEEE proposals include the possibility that radio access networks (RANs) belong to the same or different operators.

Some of the radio access network nodes are able to reconfigure their radio parameters, whose management is handled from the functional architectures (FAs). The functional architectures also provide that radio frequencies be dynamically managed, with specified rules for spectrum usage and allowed maximum powers. The FAs even enable spectrum sharing, spectrum renting and primary and secondary spectrum usages. The functional architectures bidirectionally communicate with the underlying RANs, which can belong to the same or to different operators, in order to obtain information and manage the radio spectrum efficiently. The ETSI RRS functional architecture proposes the introduction of a cognitive pilot channel (CPC) for the management of the radio context (i.e., frequencies, operating RATs, etc.).

IEEE 1900.4 standardizes the system architecture for reconfigurable radio networks and terminals, with the goal of overall radio resources optimization, including the radio spectrum. In 2011, an amendment to the IEEE 1900.4 standard, the IEEE 1900.4a, was published. IEEE 1900.4a adds to the IEEE 1900.4 functional architecture new entities for dynamic spectrum access networks in white space frequency bands. Table 7.2 resumes the possible spectrum management approaches of RRS functional architectures. Thanks to functional elements for spectrum management, like the dynamic spectrum manager (DSM) in ETSI FA and the operator spectrum (OSM) in IEEE1900.4, the proposed functional architectures lead to all the possible spectrum management approaches. The white space manager (WSM) entity in the IEEE1900.4a architecture is the entity responsible for white space management.

Table 7.2 shows that the RRS functional architectures available to all the possible spectrum management approaches.

Table 7.2 Spectrum management approaches of RRS functional architectures

System	Licensing regime	Management approach
IEEE 1900.4	Licensed	All the possible spectrum management approaches
IEEE 1900.4a	License exemption	Secondary spectrum use
ETSI FA	Licensed	All the possible spectrum management approaches

7.4 Summary

Today, licensed and unlicensed systems offer to the end user the possibility to be connected through different radio access networks (RANs), using licensed and unlicensed bands. Emerging technologies like cognitive and reconfigurable radio devices and network nodes will enable new network scenarios and new modes of spectrum usage. This will be facilitated by the development of new spectrum policies that make effective use of new technologies.

Up to now, most regulators used a command and control (CAC) model, where centralized authorities allocate the spectrum and decide its usage. Recently, new spectrum usages have been considered from the regulators, which lead to a dynamic management of the spectrum, with primary and secondary usages, with or without license.

Recent standards from IEEE and ETSI open up new scenarios of spectrum usages. IEEE 802.22 is the first system based on a cognitive radio and introduces a dynamic usage of white spaces. The functional architectures proposed by ETSI and the IEEE 1900.4 standards include functional entities for spectrum management, which facilitate the implementation of all the possible dynamic spectrum management models.

References

1. Chen, K.-C. and Ramjee Prasad, R. (2009) *Cognitive Radio Networks*, John Wiley & Sons, Ltd, Chichester.
2. Kolodzy, P. (2004) *Dynamic Spectrum Policies: Promises and Challenges*, CommLaw Conspectus.
3. http://ec.europa.eu/information_society/policy/ecomm/radio_spectrum/eu_policy/index_en.html.
4. Dixon, J. (2005) Radio spectrum management – a tutorial, 13th WWRF Meeting, 2 March 2005.
5. http://www.cept.org/ecc/.
6. http://web.oas.org/citel/en/Pages/default.aspx.
7. http://www.apt.int/.
8. McManus, P. (2009) *Environmental Regulation*, Elsevier Ltd, Australia.
9. United States General Accounting Office (2004) Better knowledge needed to take advantage of technologies that may improve spectrum efficiency. *Spectrum Management*, May.
10. Federal Communications Commission Office of Engineering and Technology Policy and Rules Division (2012) FCC Online Table of Frequency Allocations, revised on 25 May 2012.
11. http://transition.fcc.gov/oet/spectrum/table/fcctable.pdf.
12. The European table of frequency allocations and applications in the frequency range 9 kHz to 3000 GHz (ECA Table), ERC Report 25.
13. Xhao, Q. and Sadler, B.M. (2007) A survey of dynamic spectrum access. *IEEE Signal Processing Magazine*, May.
14. Saruthirathanaworakun, R. and Peha, J.M. (2010) Dynamic primary–secondary spectrum sharing with cellular systems. in *Proceedings of IEEE Crowncom*, 2010.
15. Lin, P., Jia, J., Zhang, Q. and Hamdi, M. (2010) Dynamic spectrum sharing with multiple primary and secondary users, in *Proceedings if IEEE ICC*, 2010.
16. RSPG11-392 Final, Radio Spectrum Policy Group (2011) Report on Collective Use of Spectrum (CUS) and Other Spectrum Sharing Approaches, November 2011.
17. RSPG08-244 Final, Radio Spectrum Policy Group (2008) Opinion on Aspects of a European Approach to 'Collective Use of Spectrum, 19 November 2008.
18. MacDonald, M. (2006) Study on Legal, Economic and Technical Aspects of 'Collective Use' of Spectrum in the European Community, November 2006.
19. http://ec.europa.eu/information_society/policy/ecomm/radio_spectrum_copy%281%29/topics/collective/index_en.htm.
20. Peha, J.M. (2008) Sharing Spectrum through Spectrum Policy Reform and Cognitive Radio, Department of Engineering and Public Policy, Paper 5. http://repository.cmu.edu/epp/5.

21. Niyato, D. and Hossain, E. (2008) Spectrum trading in cognitive radio networks: a market-equilibrium-based approach. *IEEE Wireless Communications*, **15** (6), December, 71–80.

22. SCF Associates Ltd (2012) Perspectives on the Value of Shared Spectrum Access: Final Report, February 2012.

23. Oliver & Ohlbaum Associates Ltd and DotEcon Ltd (2008) The Effects of a Market-Based Approach to Spectrum Management of UHF and the Impact on Digital Terrestrial Broadcasting, 27 February 2008.

24. Forge, S., Horvitz, R. and Blackman, C. (2012) Perspectives on the Value of Shared Spectrum Access, Final Report for the European Commission, 10 February 2012.

25. 1721/TNR/FR/1 (2006) Technology-Neutral Spectrum Usage Rights, Ofcom Final Report, 10 February 2006.

26. An Economic Study to Review Spectrum Pricing, Indepen, Aegis Systems and Warwick Business School, February 2004.

27. Lemstra, W. (2008) Cognitive Radio Defying Spectrum Management. A collection of four (short) contributions, in CRNI Brussels Conference, 2008.

Index